Assessment, Evaluation, and Repair of Concrete, Steel, and Offshore Structures

Assessment, Evaluation, and Repair of Concrete, Steel, and Offshore Structures

Mohamed Abdallah El-Reedy

CRC Press is an imprint of the
Taylor & Francis Group, an **informa** business

CRC Press
Taylor & Francis Group
6000 Broken Sound Parkway NW, Suite 300
Boca Raton, FL 33487-2742

© 2019 by Taylor & Francis Group, LLC
CRC Press is an imprint of Taylor & Francis Group, an Informa business

No claim to original U.S. Government works

Printed on acid-free paper

International Standard Book Number-13: 978-0-8153-6298-2 (Hardback)

This book contains information obtained from authentic and highly regarded sources. Reasonable efforts have been made to publish reliable data and information, but the author and publisher cannot assume responsibility for the validity of all materials or the consequences of their use. The authors and publishers have attempted to trace the copyright holders of all material reproduced in this publication and apologize to copyright holders if permission to publish in this form has not been obtained. If any copyright material has not been acknowledged please write and let us know so we may rectify in any future reprint.

Except as permitted under U.S. Copyright Law, no part of this book may be reprinted, reproduced, transmitted, or utilized in any form by any electronic, mechanical, or other means, now known or hereafter invented, including photocopying, microfilming, and recording, or in any information storage or retrieval system, without written permission from the publishers.

For permission to photocopy or use material electronically from this work, please access www.copyright.com (http://www.copyright.com/) or contact the Copyright Clearance Center, Inc. (CCC), 222 Rosewood Drive, Danvers, MA 01923, 978-750-8400. CCC is a not-for-profit organization that provides licenses and registration for a variety of users. For organizations that have been granted a photocopy license by the CCC, a separate system of payment has been arranged.

Trademark Notice: Product or corporate names may be trademarks or registered trademarks, and are used only for identification and explanation without intent to infringe.

Library of Congress Cataloging-in-Publication Data

Names: El-Reedy, Mohamed A. (Mohamed Abdallah) author.
Title: Assessment, evaluation, and repair of concrete, steel, and offshore structures / Mohamed Abdallah El-Reedy.
Description: First edition. | Boca Raton, FL : CRC Press/Taylor & Francis Group, 2018. | Includes bibliographical references and index.
Identifiers: LCCN 2018027139 | ISBN 9780815362982 (hardback : acid-free paper) | ISBN 9780429425455 (ebook)
Subjects: LCSH: Offshore structures—Maintenance and repair.
Classification: LCC TC1665 .E497 2018 | DDC 627.028/8—dc23
LC record available at https://lccn.loc.gov/2018027139

Visit the Taylor & Francis Web site at
http://www.taylorandfrancis.com

and the CRC Press Web site at
http://www.crcpress.com

This book is dedicated to the spirits of my mother and my father.
And to those who inspired me
To my wife and my children Maey, Hisham and Mayar.

Contents

Preface xiii
Author xv

Chapter 1 Cases of Structure Failure 1

        1.1   The Probability of Structure Failure 1
        1.2   Structure Failure 1
        1.3   Sources of Concrete Failure 8
        1.4   Structure Assessment 10
        1.5   Structure Mode of Failure 10
        Bibliography 11

Chapter 2 Project Management and Design-Induced Structure Failure 13

        2.1   Introduction 13
        2.2   Design and Structural Loading Cause Failure 13
        2.3   Lack of Design due to Interaction of Structural Load and
              Environmental Effects 14
        2.4   Poor Construction Management Cause Failure 18
        2.5   Condition Assessment and Remaining Service Life 21
        2.6   Evaluation of Reinforced Concrete Aging or Degradation
              Effects 22
        Bibliography 22

Chapter 3 Problems in Quality Management System 25

        3.1   Introduction 25
        3.2   Quality System 25
        3.3   ISO 9001 26
        3.4   Quality Manual 27
              3.4.1   Quality Plan 27
              3.4.2   Quality Control 29
              3.4.3   Quality Control and Building Failure 29
                      3.4.3.1   Submittal Data 30
                      3.4.3.2   Materials Tests 31
                      3.4.3.3   Methods of Check the Construction
                                   Activity 31
                      3.4.3.4   Material/Equipment Compliance Tests 33
                      3.4.3.5   Soils Testing 33
                      3.4.3.6   Inspection During Work-in-Progress 33
                      3.4.3.7   Pre-installation Inspection Reports 33
        3.5   Quality Assurance 34
              3.5.1   The Responsibility of the Contractor 35

		3.5.2 The Owner Responsibility 36

- 3.6 Project Quality Control in Various Stages 37
  - 3.6.1 Detailed Engineering 37
  - 3.6.2 Execution Phase 38
    - 3.6.2.1 Inspection Procedures 39
    - 3.6.2.2 Checklists 40
- 3.7 External Auditing for the Project 40
- 3.8 Operational and Maintenance Phase of the Project 42
- Bibliography 43

Chapter 4 Bad Materials Cause Failures 47

- 4.1 Introduction 47
- 4.2 Concrete Materials Test 47
  - 4.2.1 Cement 48
    - 4.2.1.1 Cement Test by The Sieve No. 170 49
    - 4.2.1.2 Define Cement Fines by Using Blaine Apparatus 50
    - 4.2.1.3 Initial and Final Setting Times of Cement Paste Using VICAT's Apparatus ... 52
    - 4.2.1.4 Compressive Strength of Cement Mortars 54
  - 4.2.2 Cement Storage 56
  - 4.2.3 Aggregate Tests 59
    - 4.2.3.1 Sieve Analysis Test 59
    - 4.2.3.2 Abrasion Resistance of Coarse Aggregates in Los Angeles Test 63
    - 4.2.3.3 Determination of Clay and Other Fine Materials in Aggregates 66
    - 4.2.3.4 Aggregate Specific Gravity Test 67
    - 4.2.3.5 Fine Aggregate Test 67
    - 4.2.3.6 Define Specific Gravity for Coarse Aggregate 68
    - 4.2.3.7 Bulk Density or Volumetric Weight Test for Aggregate 68
    - 4.2.3.8 The Percentage of Aggregate Absorption ... 69
  - 4.2.4 Mixing Water Test 70
- 4.3 Admixtures 70
  - 4.3.1 Samples for Test 72
  - 4.3.2 Chemical Tests to Verify Requirements 72
    - 4.3.2.1 Chemical Tests 72
    - 4.3.2.2 Ash Content 73
    - 4.3.2.3 Relative Density 73
    - 4.3.2.4 Define the Hydrogen Number 73
    - 4.3.2.5 Define Chloride Ion 73
  - 4.3.3 Performance Tests 74
    - 4.3.3.1 Control Mixing 74

Contents ix

          4.4   Steel Reinforcement Test 76
                  4.4.1   Weights and Measurement Test 76
                  4.4.2   Tension Test 77
          Bibliography 79

Chapter 5   Loads and Structure Failure 81

          5.1   Introduction 81
          5.2   Dead Load 82
          5.3   Live Load Characteristics 82
                  5.3.1   Previous Work on Live Load 83
                            5.3.1.1   Statistical Model for Floor Live Loads 85
                  5.3.2   Stochastic Live Load Models 87
                            5.3.2.1   Poisson Square Wave Process 88
                  5.3.3   Filtered Poisson Process 88
                  5.3.4   Analysis of the Suggested Model 90
                  5.3.5   Methodology and Calculation Procedure 93
                  5.3.6   Testing of the Suggested Model 94
          5.4   Live Loads in Different Codes 96
                  5.4.1   Comparison between Live Load for
                            Different Codes 96
                  5.4.2   Values of Live Loads and Its Factors in
                            Different Codes 97
                  5.4.3   Floor Load Reduction Factor in Different Codes 97
                  5.4.4   Comparison between Total Design Live Load
                            Values in Different Codes 100
          5.5   Delphi Method 101
          5.6   Overloads 105
                  5.6.1   Uncertainties in Calculation of Load Effects 106
                  5.6.2   Live Load Causing Failure 108
          5.7   Wind Load Statistics 108
          5.8   Earthquake Load 110
          Bibliography 113

Chapter 6   Reliability-Based Design for Structures 117

          6.1   Introduction 117
          6.2   Reliability of Reinforced Concrete Column 117
          6.3   Calculation of the Straining Actions at the Column Base 118
          6.4   Ultimate Strength of Reinforced Concrete Columns 119
                  6.4.1   Uniaxially Loaded Column 120
                  6.4.2   Biaxially Loaded Column 121
          6.5   Limit State Equation and Reliability Analysis 123
                  6.5.1   Equivalent Normal Distribution 124
          6.6   Parameters and Methodology 125
          6.7   Application on a Building 126

	6.8	Effect of Column Location	127
		6.8.1 Effect of Eccentricity	128
		6.8.2 Effect of Major Limit State Variables	130
		6.8.2.1 Effect of Concrete Strength	130
		6.8.2.2 Effect of Dead Load	132
		6.8.2.3 Effect of Steel Strength	133
	6.9	Reliability of Flexural Member	133
	6.10	Seismic Reliability Analysis of Structures	138
	6.11	Example	142
	6.12	Steel and Offshore Structure Reliability	143
	Bibliography		145
Chapter 7	Reliability of Concrete Structure Exposed to Corrosion		147
	7.1	Introduction	147
	7.2	Effect of Age on Strength of Concrete	148
		7.2.1 Researchers' Suggestions	148
		7.2.2 Code Recommendations	150
	7.3	Corrosion of Steel in Concrete	154
		7.3.1 Causes and Mechanisms of Corrosion and Corrosion Damage	154
		7.3.2 Carbonation	155
		7.3.2.1 Carbonation Transport Through Concrete	157
		7.3.2.2 Parrott's Determination of Carbonation Rates from Permeability	157
		7.3.3 Corrosion Rates	159
		7.3.4 Statistical Analysis of Initiation and Corrosion Rates	160
		7.3.5 Corrosion Effect on Spalling of Concrete	160
		7.3.6 Capacity Loss in Reinforced Concrete Columns	162
	7.4	Parametric Study for Concrete Column	162
		7.4.1 Effect of Age	163
		7.4.2 Effect of Percentage of Longitudinal Steel	164
		7.4.3 Effects of Corrosion Rate	166
		7.4.4 Effect of Initial Time of Corrosion	167
		7.4.5 Effect of Eccentricity	168
	7.5	Effect of Corrosion on the Girder	171
	7.6	Recommendations for Durable Design	177
	Bibliography		178
Chapter 8	Inspection Methodology: Visual Inspection		179
	8.1	Introduction	179
	8.2	Concrete Structure Inspection	180
		8.2.1 Collecting Data	180

Contents   xi

    8.2.2 Design Code 182
    8.2.3 Visual Inspection... 183
      8.2.3.1 Plastic Shrinkage Cracking 186
      8.2.3.2 Settlement Cracking 186
      8.2.3.3 Drying Shrinkage 187
      8.2.3.4 Thermal Stresses.................................... 188
      8.2.3.5 Alkaline Aggregate Reaction 190
      8.2.3.6 Sulfate Attack 192
    8.2.4 Environmental Considerations 193
      8.2.4.1 Chemical Attack 194
      8.2.4.2 Leaching 195
      8.2.4.3 Acid and Base Attack 195
      8.2.4.4 Steel Reinforcement Corrosion............... 196
      8.2.4.5 Salt Crystallization 200
      8.2.4.6 Freezing-and-Thawing Attack-Concrete.... 200
      8.2.4.7 Abrasion, Erosion, and Cavitation 201
      8.2.4.8 Combined Effects 201
  8.3 Steel Structure Inspection 201
    8.3.1 Visual Test... 201
  8.4 Offshore Structure 204
  Bibliography..208

Chapter 9 Inspection Methods 211

  9.1 Concrete Structures 211
    9.1.1 Core Test... 211
    9.1.2 Rebound Hammer 218
      9.1.2.1 Data Analysis...220
    9.1.3 Ultrasonic Pulse Velocity..220
    9.1.4 Load Test for Concrete Members.............................224
      9.1.4.1 I-Test Procedure 224
    9.1.5 Comparison between Different Tests 227
    9.1.6 Define Chloride Content in Harden Concrete 228
    9.1.7 Concrete Cover Measurements..................................230
    9.1.8 Measurement of Carbonation Depth 231
    9.1.9 Chlorides Test...232
      9.1.9.1 Half Cell 234
  9.2 Steel Structure 236
    9.2.1 Introduction 236
    9.2.2 Radiographic Test...236
    9.2.3 General Welding Discontinuities 236
    9.2.4 Ultrasonic Test..239
      9.2.4.1 Wave Propagation 241
    9.2.5 Reflection in Sound Wave 242
    9.2.6 Wave Interaction or Interference...............................245
    9.2.7 Penetration Test 246

		9.2.7.1	Advantages and Disadvantages of Penetrant Testing 247
		9.2.7.2	Penetrant Testing Materials 248
	9.2.8	Magnetic Particle Inspection..................................... 251	
		9.2.8.1	Magnetic Field Characteristics 252
	9.2.9	Electromagnetic Fields.. 254	
		9.2.9.1	Tools for Testing 255
9.3	Offshore Structure Inspection 257		
Bibliography..258			

Chapter 10 Repair of Concrete, Steel, and Offshore Structures.........................261

    10.1 Introduction 261
    10.2 Concrete Repair Procedure 262
          10.2.1 Removing Concrete Cover 263
                    10.2.1.1 Manual Method 265
                    10.2.1.2 Pneumatic Hammer Methods 265
                    10.2.1.3 Water Jet 265
                    10.2.1.4 Grinding Machine 267
          10.2.2 Clean Concrete Surface and Steel Reinforcement 267
          10.2.3 Concrete Repair Process 268
          10.2.4 Beam and Slab Repair Process 269
          10.2.5 Column Repair Process.. 271
          10.2.6 New Concrete Patches 274
                    10.2.6.1 Polymer Mortar... 274
                    10.2.6.2 Cement Mortar.. 274
          10.2.7 Execution Methods..275
                    10.2.7.1 Manual Method 275
                    10.2.7.2 Grouted Pre-placed Aggregate 276
                    10.2.7.3 Shotcrete 276
          10.2.8 New Methods for Strengthen Concrete Structure 279
                    10.2.8.1 Using Steel Sections 279
          10.2.9 General Precaution 284
    10.3 Steel Structure Repair 285
          10.3.1 Fatigue Crack 286
    10.4 Offshore Structure Repair 287
          10.4.1 Dry Welding at or Below Sea Surface......................287
    Bibliography..290

Index..291

Preface

Worldwide, there are many failures or complete collapses of concrete, steel, and offshore structures. If the reasons behind these failures are known and understandable, then to avoid such incidents will be much easier. In this book, the defects in project management from the feasibility stage through operation and maintenance will be fully explained. The objective of this book is to provide guidelines for structural engineers to understand the best ways of establishing functional structures that remain durable throughout its lifetime.

With my experience in lecturing many courses worldwide, I have found that the main underlying issues of failure have to do with overconfidence, and every civil or structural engineer considers that the factor of safety covers everything. However, these safety factors have a certain limit in providing the safety of structures. Therefore, it is very important to fully understand the theoretical background behind the factors of safety in different codes and standards.

Therefore, this book covers all the aspects of a project's life, such as project management pitfalls, design errors, bad materials, construction problems, and operation and maintenance that can cause the structural failure or even collapse. Also this book address the many structural failures of both onshore and offshore structures, and the underlying reasons and mechanisms of the failures will be addressed by case studies with engineering calculations. Due to structural aging, it is mandatory to perform a proper assessment and repair of the existing structure or increase its strength to achieve its full functional use. The traditional and advanced methods of assessment and inspection for concrete and steel structure are presented, along with the best ways to predict the structural lifetime for onshore and offshore structures.

Dr. Mohamed Abdallah El-Reedy
elreedyma@gmail.com
www.elreedyman.tk

Author

Mohamed Abdallah El-Reedy, PhD, is a practicing structural engineer. His main area of research is the reliability of concrete and steel structures. He has provided consulting services to different engineering companies and oil and gas industries in Egypt and to international companies including the International Egyptian Oil Company (IEOC) and British Petroleum (BP). Moreover, he provides concrete and steel structure design packages for residential buildings, warehouses, telecommunication towers, and electrical projects of WorleyParsons Egypt. He has participated in liquefied natural gas and natural gas liquid projects with international engineering firms. Currently, Dr. El-Reedy is responsible for reliability, inspection, and maintenance strategies for onshore concrete structures and offshore steel structure platforms. He has managed these tasks for one hundred of these structures in the Gulf of Suez in the Red Sea.

Dr. El-Reedy has consulted with and trained executives at many organizations, including the Arabian American Oil Company (ARAMCO), British Petroleum (BP), Apachi, Abu Dhabi Marine Operating Company (ADMA), the Abu Dhabi National Oil Company and King Saudi's Interior Ministry, Qatar Telecom, Egyptian General Petroleum Corporation, Saudi Arabia Basic Industries Corporation, the Kuwait Petroleum Corporation, Qatar Petrochemical Company (QAPCO) and PETRONAS, Malaysia, and PTT, Thailand. He has taught technical courses about repair and maintenance for reinforced concrete structures and advanced materials worldwide.

Dr. El-Reedy has written numerous publications and presented many papers at local and international conferences sponsored by the American Society of Civil Engineers, the American Society of Mechanical Engineers, the American Concrete Institute, the American Society for Testing and Materials, and the American Petroleum Institute. He has published many research papers in international technical journals and has authored seven books about total quality management, quality management and quality assurance, economic management for engineering projects, and repair and protection of reinforced concrete structures, advance materials, onshore projects in oil and gas projects, offshore structure and concrete structure reliability. He earned a bachelor's degree from Cairo University in 1990, a master's degree in 1995, and a PhD from Cairo University in 2000.

1 Cases of Structure Failure

The failure of any structure has a general meaning that it is not doing its function. A simple example is a crack or deterioration that prevents the user from using the structure as he doubts about his safety. The crises in the case of the collapse of the structure are considered to be the complete failure of the structure. All the codes and standards concerning about preventing the collapse of the building are maintained well the structure serviceability within the permissible limit.

The failure is happened under main circumstances, but in general in case of occurring in a time period of the structure life there is a maximum load and minimum strength happened then the failure will occur.

1.1 THE PROBABILITY OF STRUCTURE FAILURE

We find that loads when calculated from codes and standards are taken as one value, but in fact the load is presented by a probability distribution. When calculating the member resistance capacity from the code equation, it provides one value. Therefore, our calculation is called a "deterministic calculation," but in the case of taking every parameter as variable presented by a probability distribution with mean value and coefficient of variation, the member resistance capacity will be presented by a probability distribution and in this case the calculation is called probabilistic calculation.

Figure 1.1, shows that the probability of structure failure occurs in the zone of intersection between the two curves of load and resistance. The failure of structure may be a partial collapse, total collapse, or member failure. This failure happens as the load value is at its highest value and the strength is at its lowest value due to deterioration, poor quality control, or incorrect design.

From a practical point of view, the load increases due to a change in the mode of operation or there are loads accumulating at the same time, for example, collecting all furniture in one room during painting or the load increases in case of parties.

The strength is low usually due to poor quality control during construction or concrete deterioration due to corrosion in the steel bars and spalling concrete cover or cracks in concrete structure member in old buildings.

1.2 STRUCTURE FAILURE

There are many structure failures worldwide. Here we just collect the famous structural failures and try to find the reason for these failures. The following are different cases studies.

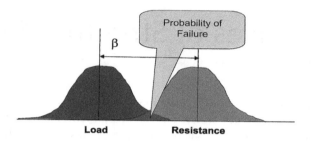

FIGURE 1.1 Probability of failure diagram.

Case Study 1

Figure 1.2 shows the inclination of a building in Alexandria in 2017. The major cause of the structural failure of the building is defective design and poor dealing in design with weak soil. Therefore, it is required to do a pile foundation. However, as a cheap option the engineer uses the raft foundation to increase the number of the floors. An initial investigation found that this building collapsed a few years after the completion of construction.

Case Study 2

Ten years ago, the Interstate 35W bridge over the Mississippi River in downtown Minneapolis collapsed, sending cars, trucks, and even a school bus that were

FIGURE 1.2 Building collapse at Alexandria.

crawling over it in bumper-to-bumper rush hour traffic plummeting into the river below and onto the rocky shore, leaving 13 people dead and 145 people injured. Many of them were seriously injured.

The bridge collapse sparked immediate calls in Minnesota and across the country invest big in repairing and replacing the nation's aging and crumbling infrastructure.

A decade later, experts say there have been some improvements, but there are still tens of thousands of bridges nationwide that need to be fixed or replaced.

In the immediate aftermath of the 35W bridge collapse, the Minnesota Department of Transportation came under intense scrutiny. The interstate highway bridge had been classified as structurally deficient, meaning that it was aging and in need of repair. In fact, some repair work was going on when it fell. The bridge was also rated as fracture critical, meaning that the failure of just one vital component could cause the whole bridge to collapse.

According to ASCE team investigation in 2007, the thickness of gusset and the thickness of the side wall of the upper chords were designed proportional to the bending moment solution of a one-dimensional influence line analysis.

This fact reveals that the undersized gusset plates are the consequence of a bias toward a "one-dimensional model" in the original design that did not give sufficient consideration to the effects of the forces from diagonal truss members.

Although the bridge's truss structure was appropriately designed, the design of the node that connected the floor members to the main truss-frame was inadequate to effectively distribute live and dead loads. Consequently, the local redundancy provided by the truss-cells was significantly reduced.

A three-dimensional, nonlinear, finite-element, computation-based load rating indicates that some of the gusset plates had almost reached their yield limit when the bridge experienced the design load condition. The bridge was sustained by the additional safety margin provided by the ultimate strength of the ductile steel that comprised the gusset plates(see ASCE 2010).

Case Study 3

This failure was happened in Israel in 2001 during a wedding ceremony leaving at least 24 people dead, including a 3-year-old boy. The officials estimated that over 300 people injured and countless others trapped beneath the rubble of the collapsed building.

Sarne, who for years warned about the dangers of Pal-Kal both before and after the Versailles tragedy, now actually sounds more relaxed: "A building won't collapse due to Pal-Kal construction alone, but due to a combination of circumstances, such as in the Versailles disaster," he says. "Pal-Kal doesn't fall by itself, only with help. The proof is that hundreds of Pal-Kal buildings have been identified yet they aren't constantly collapsing. They may not be up to standard and do pose a certain risk, but they aren't collapsing." (see Haaretz 2011).

So the new technique in design and construction is the main reason of this failure so it needs more verifications before use any new technology. If you can see the slab failure it is complete a punching failure of the slab as per the yield line failure.

Case Study 4

The bridge in Florida was collapsed in March 2018 during the writing of this book. The pedestrian bridge that collapsed at Florida International University in Miami had been put into place 5 days earlier and was being built using a method called accelerated bridge construction. The main, 53 m (174 feet) walkway of the bridge was assembled next to Southwest Eighth Street, a major thoroughfare that separates the university campus from the city of Sweetwater. Two days before the bridge collapse, the walkway was rotated and moved to rest on two supports that were built on either side of the street. The engineer reported cracks on the north end of the bridge 2 days before it collapsed. The main design of the final bridge, which was proposed to complete in 2019, would have added a tower and cable stays as the main support for the structure. As a result, the collapse crushed cars and killed at least six people. In this scenario, the main reason is due to engineering errors but for fast-track construction the error during construction is highly the main factor.

Case Study 5

Figure 1.3 shows a bridge collapse in Egypt in 2014 due to lack of maintenance of the bridge. As it is mature structure and there is no clear maintenance plan, therefore the concrete strength will be decrease with time, which causes a failure in

FIGURE 1.3 Bridge collapse in Egypt.

the case of occurring the maximum load event on the structure or it can be cause due to the maximum applied load more than that was considered in the design.

Case Study 6

The death toll from the collapse of a 12-storey residential building in Egypt's Mediterranean city of Alexandria in 2007.

The building was home to between 40 and 50 people but the accident happened in the morning after many residents had left for work or school.

By investigation they found that the local authorities had ordered the removal of the building's top two floors in 1995 because they contravened building laws, but the order was not carried out.

The officials estimated that 25–30 people were still beneath the debris.

The investigation revealed that construction workers had been doing repairs on the first floor when the building suddenly tilted and collapsed.

The sources added that the owner had built a seven-storey building in 1982 without a permit and later secured a license, but had then illegally built an additional five stories.

The authorities had issued at least two orders for renovation work to be done on the building, but it was not carried out.

Streets leading to the site of the collapse were sealed off. Buildings on either side of the collapsed block suffered structural damage and were evacuated (see The Age 2007).

Case Study 7

A 13-storey apartment building buried and killed a worker in China. This collapse was happened due to bad design of the piles.

In this failure, all the structure system above ground was good and was put in a horizontal position. However, the major cause of collapse was water in soil since the building was situated near to a river. The main question arose is why did this building collapse and the surrounded buildings did not. As our engineering sense and by close investigation to the collapse, one can find that the reason there is a low strength for foundation and soil interaction design capacity and high load. In this case the dealing with soil is very complicated as may be any factor is affecting the soil will produce dramatic decrease to its strength.

Case Study 8

Figures 1.4 and 1.5 show the settlement problem in buildings. There are cracks for walls and different buildings in whole new city in Saudi Arabia. The settlement is a usual reason which causes all these problems.

This is a new area which was built on the desert, this case is very critical rather a new building in the downtown as in the urban area the nature of soil is

FIGURE 1.4 Cracks on the wall in Saudi Arabia.

FIGURE 1.5 Wall inclination and cracks due to settlement.

known clearly due to many soil boring and soil investigation were done which let the designers have a confidence about the nature of the soil. On the other hand, in the case of the new city we need to be afraid and more care about the soil which request many soil boring and soil tests to define exactly the condition of the soil, in particular, from the point of view of settlement.

These figures present the settlement for different buildings in the same area. It is worth to mention that in sandy soil the settlement effect will present after finishing the construction rather than the clay soil as the settlement will present from 8 to 10 years after construction.

Case Study 9

Figure 1.6 shows the collapse of Dubai building during construction. The foundation of this building during the construction was very good and the quality of the work was up to the international standard.

Cases of Structure Failure 7

FIGURE 1.6 Dubai building during construction.

FIGURE 1.7 Dubai building collapse.

FIGURE 1.8 Dubai building collapse.

The failure was happened in 2007 after the completion of the construction work. The building was suddenly collapsed before starting the construction work. It is worth to mention that this structural failure of the building was occurred due to high load only. Thanks to God no one was injured or killed in this accident. As can be seen from Figures 1.7 and 1.8, the slab is a composite section as a beam from steel and the slab from concrete. The steel reaches to plasticity as seen in the figure and there is a tearing in the steel connection. Hence, this failure is happened due to faulty or defective design, which is the most common factor for any steel structure failure.

Case Study 10. Offshore failure

Figure 1.9 presents not a complete collapse but a very serious condition due to high degradation to the structure member. This offshore jacket reaches to this problem as this jacket is supporting the flare so the continuous inspection is not achievable.

FIGURE 1.9 Deterioration on offshore jacket.

Case Study 11

Figure 1.10 presents a high deterioration of the cantilever of the harbor jetty. This is due to high load for moving crane in addition to lower strength due to corrosion of the steel bars. As a precaution to this condition, it is prohibited to move any truck over this jetty, so it is considered as a failure to its objective. This situation is remaining until we do a complete repair toe the jetty.

1.3 SOURCES OF CONCRETE FAILURE

The corrosion is not the only source of failure. There are many other sources that can cause determination on the reinforced concrete structures. This must be kept in mind and understood well when an inspection is undertaken. These sources of failure are as follows:

Cases of Structure Failure

FIGURE 1.10 Harbor jetty failure.

I. Unsuitable materials
- Unsound aggregate
- Reactive aggregate
- Contaminated aggregate
- Use wrong type of cement
- Cement manufacturer error
- Wrong type admixture
- Substandrad admixture
- Contaminated admixture
- Organic contaminated water
- Chemical contaminated water
- Wrong kind of reinforcement
- Size error of steel bars

II. Improper workmanship
- Faulty design
- Incorrect concrete mixture (low or high cement content, and incorrect admixture dosage)
- Unstable formwork
- Misplaced reinforcement
- Error in handling and placing concrete (segregation, bad placing, and inadequate compact)
- Curing incomplete

III. Environmental
- Soil alkali
- Sea water or sewage
- Acid industry
- Freezing and thawing

IV. Structural
- Load exceed design
- Accident as ballast load or dropped object
- Earthquake load

1.4 STRUCTURE ASSESSMENT

The tests that have been clarified in the preceding sections are only tools to assist in the process of evaluating the structure. The main factor in this process is the experience of the engineer who is examining the structure.

After collecting all the required data, start the evaluation process by performing a structural analysis for the critical member or the whole structure to calculate bending moment, shearing force, normal force on the reinforced concrete member, and define the concrete structure member capability to carry the predicting apply load during its remaining service life. In most cases, the linear structure analysis in scope of elastic theory will be applied as it is familiar with all the engineers, but you should know that it is conservative. The nonlinear structure analysis with the plasticity theory is complicated and is not familiar to the major of engineers; however, these days most structure analysis software package has a pushover module, but practically it is most useful to apply in steel structure rather than the reinforced concrete structure where it is not easy. There are many research studies covering this issue, but it is not applied easily in practice. However, the evaluation of the deteriorated structure is mainly dependent on the experience of the engineer who is performing this evaluation.

The evaluation of the building should determine whether the structure is subject to repair or not, if the situation entirely bad, or if the cost of repairs is nearly the same as the demolition of building and its reconstruction.

Therefore, we note that the evaluation of the building is a vital and critical to the decision process as the decision-making will be based on an assessment of the building in terms of safety and repair cost.

Therefore, the repair scope of work is based on the accuracy of the test results and previous experience of skilled engineers who conducted these investigations.

When determining the method of repair and the determination of the necessary cost value of the repair process, one must not forget to keep in mind the methods that require protecting the structure from corrosion after repair.

1.5 STRUCTURE MODE OF FAILURE

The most common cause of failure found in the investigated cases is poor design or lack of strength design is the lowest probability as 15% of the structural failures are caused by design errors, 60% by construction error, and the remaining 25% by lack of maintenance or bad operations.

The design is lower percentage of causing the structure failure as the design engineers are not working under pressure and have a time to do many review cycles so the error will be based on the project manager and the technical team.

The percentage of lack of maintenance increases with time with increasing the number of mature structure and the quality control level for most of the project. As of now, most of the operator or owner of the residential building or industrial building is still working by run-to-fail approach. This approach is not a good approach as the cost is very high and also there is a risk of corrective maintenance failure. The new approach is preventive than proactive. Now we are working by risk-based inspection but this is implemented in very critical structure like the offshore platform. However, the normal building is still working by run-to-fail approach.

Figure 1.1 is a master key for the whole book. In the case of deteriorated structure, the strength will be reduced in the case of concrete structure due to the corrosion of the steel bars. Therefore, the strength probability curve will move toward the left, and the probability of failure will increase. Thus, the probability of failure is function of time specific for steel structure but it needs some calculation for concrete structure as concrete strength gains some strength will gain with time to some limit but depends on the environmental condition which will be discussed in detail in the following chapters.

BIBLIOGRAPHY

ASCE. Journal of Bridge Engineering/Volume 15 Issue 5—September 2010. http://ascelibrary.org/doi/pdf/10.1061/(ASCE)BE.1943-5592.0000090.

ASCE. Report. August 1, 2007. www.npr.org/2017/08/01/540669701/10-years-after-bridge-collapse-america-is-still-crumbling.

Haaretz. Journal article. May 23, 2011. www.haaretz.com/print-edition/business/waiting-for-the-floor-to-collapse-1.363425.

The age. Journal article. December 5, 2007. www.theage.com.au/news/world/egyptian-building-collapse-toll-rises/2007/12/25/1198344998335.html.

2 Project Management and Design-Induced Structure Failure

2.1 INTRODUCTION

There are many factors that directly or indirectly affect the structure failure. One of the biggest failures of any project is not based on cost or to extend the time schedule or maintain safety records, but it is to have a safe structure sustainable to its life time.

Most project management professional who work well in this area considered that these failures are totally the responsibility of the technical people only. However, in fact, it is the responsibility to all staff even if the project manager has the power to transferee the problem to the technical people only. Theoretical as the project manager will not be in jail according to law but actually your reputation will be destroyed considering that in case of structure failure, it will be podcast over all media as the story of fail is very famous and interesting. Every one like to speak about it rather than your success story for other projects you did before.

The project manager leads the team and acts a maestro, coordinating, and facilitating all the different resources needed to take a project from concept to completion. Therefore, the project manager is ethically bound to prevent any potential risk responsible for a structural failure. The big question is that we can agree that his responsibility for the short period for 1 or 10 years after construction, but what about after long time it will out of his hand. No sir if he does a sustainable design provide and easy and cost effect maintenance scenario according to the environment condition in this case really the sustain project is not a dream.

2.2 DESIGN AND STRUCTURAL LOADING CAUSE FAILURE

Designers of a new project involving concrete structures address service life by defining several critical concrete parameters, such as *w/c ratio*, admixtures, reinforcement protection (cover or use of epoxy coating), and curing methods. The designer also authenticates various serviceability criteria, such as deflection and crack width. Other factors to promote durability are also addressed at this stage (see, e.g., design, drainage to minimize moisture accumulation, and joint details).

To implement a design to provide a durable reinforced concrete structure considering all the above parameters that affect the service life may be a costly option.

Improper selection of the exposure rating can lead to a more permeable concrete resulting in faster chloride penetration and diminished service life.

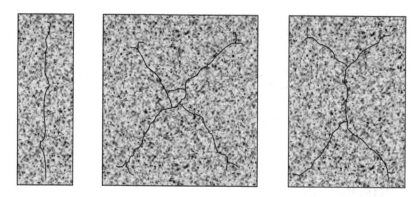

FIGURE 2.1 Cracks due to increased load on the slab.

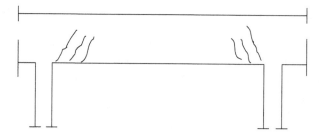

FIGURE 2.2 Cracks due to increased shear stress.

FIGURE 2.3 Cracks due to increase bending stress.

Figures 2.1–2.3 show the effect of increasing load on the reinforced concrete structure on different members due to lack of management of change policy for building use.

2.3 LACK OF DESIGN DUE TO INTERACTION OF STRUCTURAL LOAD AND ENVIRONMENTAL EFFECTS

The load may be increased due to poor design or during operation there is no management of change procedure in the case of industrial building. Management of change starts with ISO 9001 as a concept and shall be applied by the owner. On the

Project Management and Structure Failure

residential building, the management of change totally depends on the government rules and laws. In some countries, it is not allowed to convert any buildings from one type to another such as converting residential buildings to hospitals, etc. However, in the developing countries there are no any hard-and-fast rules for such constructions. In this case, the experienced designer is very important to do a reliable design considering the country culture, as he can be more conservative rather than the design code that he follows. However, some designers consider one floor building more than the client required to have a more safety for the column in calculating load.

On the other hand, the same cracks, as shown in Figure 2.1, present the cracks on the concrete slab due to an increase in load over the design or the capacity of the slab is not capable to carry the load. The crack will be longitudinal in the case of rectangular slab with a length equal or higher than twice the width. The other two crack shapes present a rectangular or square concrete slab.

Figure 2.2 presents the cracks on the side of the beam due to an increase in a shear stress due to poor design. Figure 2.3 presents the shape of cracks due to increase in the bending stress. The case of compression failure in reinforced concrete beam will occurred if the beam was design by over reinforced and the load was increase, so the shape of cracks will be as shown in Figure 2.4.

In case of increase the load on the column due to many circumstances will cause some cracks as a sign of increase the load.

Figures 2.1–2.5 show the effect of increasing load on the reinforced concrete structure in different members.

On the other hand, there are some cracks appearing on the masonry bricks due to an increase in the load or a decrease on the member capacity (see Figure 2.6). These cracks are occurred due to beam deflection. Figure 2.7 shows the diagonal cracks due to the settlements in the foundation.

According to Jacob (1965), actions to eliminate or minimize any adverse effects resulting from environmental factors and designing structural components to withstand the loads anticipated do not necessarily provide a means to predict the service life of a structure under actual field conditions. The load-carrying capacity of a structure is directly related to the integrity of the main members during its service life. Therefore, a quantitative measure of the changes in the concrete integrity with time provides a means to estimate the service life of a structure.

Several researchers have tried quantifying the environmentally induced changes by measuring the physical properties of concrete specimens after subjecting them to

FIGURE 2.4 Cracks due to compression failure.

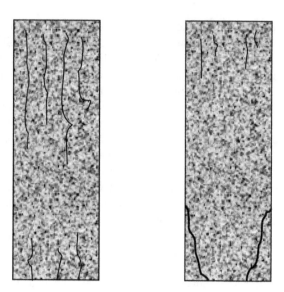

FIGURE 2.5 Cracks due to increase column load.

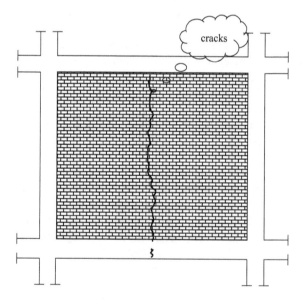

FIGURE 2.6 Cracks in masonry wall due to beam deflection.

various combinations of load and exposure (Woods 1968, Sturrup and Clendenning 1969, Gerwick 1981). Most of the physical and mechanical properties are determined using relatively small specimens fabricated in the laboratory or sampled from structures. The properties measured reflect the condition of the specimens tested rather than the structure in the field because the test specimen and structure often

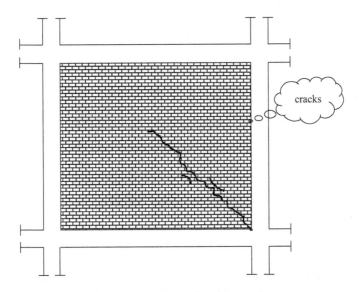

FIGURE 2.7 Inclined cracks in masonry wall due to settlement.

are exposed to somewhat different environments. Quantifying the influence of environmental effects on the ability of the structure to resist the applied loads and to determine the rate of degradation as a result is a complex issue. The application of laboratory results to an actual structure to predict its response under a particular external influence requires engineering interpretation.

Neville (1991), Sturrup et al. (1987), Avram (1981), and Price (1951) studied and reported the effect of external influences, such as exposure or curing conditions, on the changes in concrete properties.

In general, the environmental conditions affect the deterioration of reinforced concrete structure due to the following two major factors:

- The presence of moisture and the transport mechanism controlling movement of moisture or aggressive agents (gas or liquid) within the concrete.
- The transport mechanism is controlled by the microstructure of the concrete, which in turn is a function of several other factors such as age, curing, and constituents. The microstructure comprises a network of pores and cracks in the concrete.

According to the previous studied, it is revealed that the pore characteristics are a function of the original quality of the concrete, while cracking occurs in the concrete due to external loading as well as internal stresses. Ingress of aggressive agents is more likely to occur in the cracked region of the concrete than in an uncracked area. It is, therefore, possible that cracks occurring due to the service exposures affect the remaining service life of the concrete.

A quantitative measurement of the concrete microstructure can be considered in terms of permeability. Most of the techniques for measuring concrete permeability are comparative and this is not a standard test method. Kropp and Hilsdorf (1995) presented standard methods that have also been developed for testing nonsteady-state water flow. Extensive development work is needed before such techniques can be applied to predict the remaining service life of a structure. Ludwig (1980) studied a periodic measurement of water, gas, chloride permeability, or depth of carbonation and found means of quantifying the progressive change in the microstructure of concrete in service. Temper (1932) presented a type of approach that has been used to predict the service life of dams subject to leaching of the cement paste by percolating soft water.

2.4 POOR CONSTRUCTION MANAGEMENT CAUSE FAILURE

Most often, the construction methods employed meet both the intent and the details of the plans and specifications. In some instances, however, the intent of the plans and specifications are not met, either through misunderstanding, error, neglect, or intentional misrepresentation. With the exception of intentional misrepresentation, each of these conditions can be discussed through an examination of the construction process. Service-life impairment can result during any of the four stages of construction: material procurement and qualification, initial fabrication, finishing and curing, and sequential construction. With the exception of material procurement and qualification, each stage and the corresponding service-life impacts are discussed as follows.

The fabrication is defined as all the construction up to and including placement of the concrete. This work incorporates soil/subgrade preparation and form placement; reinforcement placement; and concrete material procurement, batching, mixing, delivery, and placement.

Therefore, improper soil/subgrade preparation can lead to excessive or differential settlement in the wooden or steel forms. This can result in misalignment of components or concrete cracking. Initial preparation and placement of the formwork not only establishes the gross dimensions of the structure but also influences certain details of reinforcement and structure performance.

The problems due to steel detailing are shown in Figure 2.8. In some cases with poor quality control, there will be a deviation from the construction drawings to that was constructed on site and, in some cases, there is a human error in putting the correct size of the steel bars.

In some cases, the beam was not constructed on its location outlined in the drawings or the deviation is higher than the allowable or there is higher deviation on the location of the above column. Therefore, in this case there will be an eccentricity on the column and the cracks are shown in Figure 2.9.

Concrete can be batched either on the project site or in a batch plant and transported to the site, which is most traditional these days. Activities influencing the service-life performance include batching errors, improper equipment operation, or improper preparation.

Project Management and Structure Failure

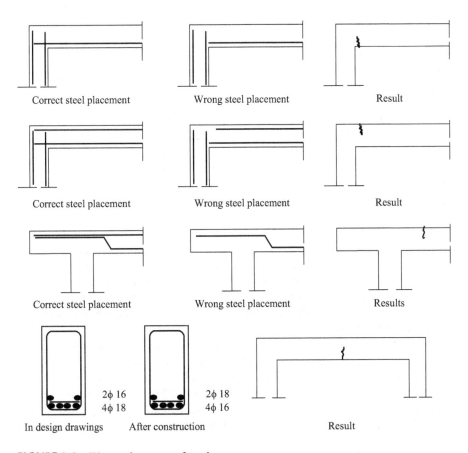

FIGURE 2.8 Wrong placement of steel.

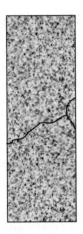

FIGURE 2.9 Cracks due to eccentricity.

In the batch plant, the mixing of the concrete component and its percentage according to the design mix is performed through computer-controlled weight and batching equipment. Sources of error are lack of equipment calibration or incorrect mixture selection. Routine maintenance and calibration of the equipment ensures proper batching. Because plants typically have tens to hundreds of mixture proportions, batching the wrong mixture is a possibility. Errors, such as omission of air-entraining admixture, inclusion of excessive water, or low cement content, are likely to have the greatest impact on service life.

Equipment preparation is the source of more subtle effects.

For example, wash water retained in the drum of a transit mix truck mixes with newly batched concrete to result in a higher *w/c* ratio than specified. This effect is cumulatively deleterious to service life through lower strength, increased shrinkage cracking, or higher permeability.

Ambient temperature, transit time, and admixture control are some of the factors controlling the mixture quality in the delivery process. Workability at the time of delivery, as measured by the slump, is also a long-term service-life issue. Low slump is often increased by adding water at the site. If the total water does not exceed that specified, concrete integrity and service life will not be reduced. If the additional water increases the total available water above that specified, then the increased *w/c* can compromise the service life.

On the other hand, the proper placement, including consolidation and screeding, is important to the service life of concrete structures. The lack of proper consolidation leads to low strength, increased permeability, loss of bond, and loss of shear or flexural capacity. These, in turn, diminish service life by accelerating the response to corrosive environments, increasing deflections, or contributing to premature failures. The following two examples illustrate how this service-life impairment can occur.

Finishing and curing. Improper finishing or curing leads to premature deterioration of the concrete and reduction of service life (e.g., production of a porous and abrasive cover concrete).

Sequential construction. Reinforced concrete structures are seldom completed in a single construction activity. Complementary or sequential construction can adversely affect the service life of the structure if not properly accomplished.

In the case of multiple-story buildings, shoring is used to support the formwork for placing concrete on the next floor. The normal practice is to remove the shoring when the form is removed and then to reshore until the concrete has gained sufficient strength to carry the construction loads. The removal of the premature form leads to cracking of the affected component. The cracking reduces the stiffness of the slab, and increases the initial deflections and the subsequent creep deflections. Even when the concrete eventually gains its full strength, the cracked member has greater deflection than a comparable uncracked member, and can be more vulnerable to ingress of hostile environments.

Joints are placed in buildings and bridges to accommodate contraction and expansion of the structure due to creep, shrinkage, and temperature. Improperly designed or installed joints can lead to excessive cracking, joint failure, moisture penetration into the structure, and maintenance problems.

2.5 CONDITION ASSESSMENT AND REMAINING SERVICE LIFE

The key factors in predicting service life and in maintaining the capability of reinforced concrete structures to meet their operational requirements include the detection and assessment of the magnitude and rate of occurrence of environmental factor-related degradation. It is desirable to have an evaluation methodology that, given the required data, provides the procedures for performing both a current condition assessment and certifying future performance. Such a methodology would integrate service history, material and geometry characteristics, current damage, structural analyses, and a comprehensive degradation model. For completeness, the methodology should also include the capability to evaluate the role of maintenance in extending usable life or structural reliability.

Verification that the structural condition is as depicted in the construction documents, such as drawings, determination of physical condition, quantification of applied loads, and examination of any degradation, is important. The questions faced in predicting service life include establishing how much data should be accumulated, the desired accuracy of the predictions, available budgets for the predictive effort, as well as subsequent levels of inspection, maintenance, and repair.

1. **Conformance of structure to original design**
   - Documentation review
   - Preliminary site visit
   - Visual inspection for compliance with construction documents
   - Pachometer (covermeter) survey to locate and characterize steel reinforcement (e.g., size and spacing)
   - Preliminary analysis
2. **Inspection for presence of degradation**
   - Visual inspection
   - Crack survey
   - Delamination/spall survey
   - Chloride survey
   - Carbonation survey
   - Sample removal
3. **Laboratory testing**
   - Petrographic studies (e.g., air content, air-void distribution, unstable aggregates, types of distress, and estimation of w/cm)
   - Chemical studies (e.g., chemical constituents of cementitious materials, pH, presence of chemical admixtures, and characteristics of paste and aggregates)
   - Concrete and steel reinforcement material properties (e.g., strength and modulus of elasticity)
4. **Degradation assessment**
   - Current-versus-specified material properties
   - Concrete absorption and permeability (relative)
   - Concrete cover (e.g., cores, or pachometer or covermeter measurements)

- Presence of excessive concrete crack widths, spalling, or delamination
- Depth of chloride penetration and carbonation
- Steel reinforcement corrosion activity (e.g., half-cell potential measurements, and galvanostatic pulse, four-electrode, and corrosion probes
- Environmental factors (e.g., presence of moisture, chlorides, and sulfates)

5. **Structural reanalyzes for current conditions**
   - Reanalyzes for typical dead and live loads
   - Examination of demands from other loads (e.g., seismic and wind)

2.6 EVALUATION OF REINFORCED CONCRETE AGING OR DEGRADATION EFFECTS

Testing is also undertaken for the verification of models, materials, and environmental parameters used for calculating the service life in the design phase. The validated or improved models are then used for optimization of the building operation and maintenance.

The physical condition and functioning of structural materials and components are the main factors to measure the performance of a building structure. According to Murphy (1984), the tests are conducted on reinforced concrete to assess performance of the structure.

Noncompliance of materials properties with specifications; Inadequacies in placing, compacting, or curing of concrete; Damage resulting from overload, fatigue, freezing and thawing, abrasion, chemical attack, fire, explosion, or other environmental factors; or Concern about the capacity of the structure. Examination of demands from other loads (e.g., seismic and wind).

The main objective to define the ability of a reinforced concrete structure to meet its functional and performance requirements over an extended period of time is largely dependent on the durability of its components. Techniques for the detection of concrete component degradation should address the concrete, steel reinforcement, and anchorage embedments.

BIBLIOGRAPHY

AASHTO T260-84, Standard method of sampling and testing for total chloride ion in concrete ratio materials, American Association of State Highway Transportation Officers, Washington, DC, 1984.

ACI 228-89-R1, In-place methods for determination of strength of concrete.

ACI 318-97, Standard building code requirements for reinforced concrete, Detroit, 1997.

Alldred, J.C., Quantifying the losses in cover-meter accuracy due to congestion of reinforcement, *Proceedings of the Fifth International Conference on structural Faults and Repair*, Vol. 2, Engineering Techniques Press, Edinburgh, pp. 125–30, 1993.

American Concrete Institute, Corrosion of metals in concrete, Report by ACI Committee 222. ACI222R-89, American Concrete Institute, Detroit, MI, 1990.

ASTM C1152, Test method for acid-soluble chloride in mortar and concrete.

ASTM C1202, Test method for electrical indication of concrete's ability to resist chloride.

ASTM C341, Test method for length change of drilled or sawed specimens of hydraulic-cement mortar and concrete.

Project Management and Structure Failure

ASTM C42-90m, Standard test method for obtaining strength and testing drilled cores and sawed beams of concrete.

ASTM C597, Standard test method for pulse velocity through concrete.

ASTM C682, Standard test method for evaluation of frost resistance of coarse aggregates in air-entrained concrete by critical dilation procedures.

ASTM C805, Test method for rebound number in concrete.

ASTM C876, Standard test method for half cell potentials of reinforcing steel in concrete.

ASTM C876, Standard test method for half-cell potentials of uncoated reinforcing steel in concrete, American Society for Testing and Materials, Philadelphia, PA, 1991.

ASTM D1411-82, Standard test methods for water soluble chlorides present as admixes in graded aggregate road mixes, American Society for Testing and Materials, Philadelphia, PA, 1982.

Avram, C., 1981, *Concrete Strength and Strain*, Elsevier Scientific Publishing Co., New York.

British Standard Institution-BS 1881-1983-Part 120, Method for determination of the compressive strength of concrete cores.

Broomfield, J.P., Langford, P.E., and McAnoy, R., Cathodic protection for reinforced concrete: its application to buildings and marine structures, in Corrosion of Metals in Concrete, *Proceedings of Corrosion/87 Symposium*, Paper 142, NACE, Houston, TX, pp. 222–325, 1987.

Broomfield, J.P., Rodriguez, J., Ortega, L.M., and Garcia, A.M., Corrosion rate measurement and life prediction for reinforced concrete structures, *Proceeding of Structural Faults and repair—93*, Vol. 2, Engineering Technical Press, University of Edinburgh, pp. 155–64, 1993.

BS 1881, Testing Concrete: Part 5: Methods for testing hardened concrete for other than strength.

BS 1881, Part 6-1971, Methods of testing Concrete, Analysis of Hardened Concrete.

BS 8110, Structural user of Concrete-Part 1:1985 Code of Practice for Design and Construction, London.

Bungey, J.H. (ed.), Non-destructive testing in Civil Engineering, *International Conference by The British Institute of Non-Destructive Testing, Liverpool University*, 1993.

Concrete Society, Technical report No. 1 including addendum (1987). Concrete core testing for strength cement and concrete association, Wexham spring, sloogh SL3 6PL.

Concrete Society, Repair of concrete damaged by reinforcement corrosion, Technical Report No.26., 1984.

El-Reedy, M.A., 2009, *Advanced Materials and Techniques for Reinforced Concrete Structures*, CRC press, Boca Raton, FL.

Gerwick, Jr., B.C., 1981, High-amplitude low-cycle fatigue in concrete sea structures, *PCI Journal*, Vol. 26, No. 5, pp. 82–97.

Jacob, F., 1965, *Lessons from Failures of Concrete Structures*, Monograph No. 1, American Concrete Institute, Farmington Hills, MI.

Kropp, J., and Hilsdorf, H.K., Performance criteria for concrete durability, TC-116-PCD, International Union of Testing and Research Laboratories for Materials and Structures (RILEM), E&FN Spon, Cachan Cedex, France, 1995.

Malhotra, V.M., and Carino, N.J., 1991, *Handbook of Nondestructive Testing of Concrete*, CRC Press, Boca Raton, FL.

Malhotra, V.M., (ed.), 1984, *In Situ/Nondestructive Testing of Concrete, SP–82*, American Concrete Institute, Farmington Hills, MI, 829 pp.

Murphy, W.E., 1984, Interpretation of tests on strength of concrete in structures, In: *Situ/Nondestructive Testing of Concrete, SP-82*, Malhotra, V.M. (ed.), American Concrete Institute, Farmington Hills, MI, pp. 377–392.

Naus, D.J., and Oland, C.B., Structural aging program technical progress report for period Jan. 1, 1993, to June 30, 1994, ORNL/NRC/LTR-94/21, Martin Marietta Energy Systems, Oak Ridge National Laboratory, Oak Ridge, TN., Nov, 1994.

Neville, A., 1991, *Properties of Concrete*, John Wiley & Sons, Inc., New York.

Parrott, L.J., 1987, *A Review of Carbonation in Reinforced Concrete*, a review carried out by C&CA under a BRE contract, British Cement Association, Slough.

Penetration.

Price, W.H., 1951, Factors influencing concrete strength, *ACI Journal, Proceedings*, Vol. 47, No. 2, pp. 417–432.

Rewerts, T.L., 1985, Safety requirements and evaluation of existing buildings, *Concrete International*, Vol. 7, No. 4, pp. 50–55.

Sturrup, V.R., and Clendenning, T.G., The evaluation of concrete by outdoor exposure, Highway Research Record HRR-268, Washington, DC, 1969.

Sturrup, V.R., Hooton, R., Mukherjee, P., and Carmichael, T., Evaluation and prediction of durability—Ontario hydro's experience, concrete durability, *Proceedings of the Katherine and Bryant Mather International Symposium, SP-100*, Scanlon, J.M. (ed.), American Concrete Institute, Farmington Hills, MI, pp. 1121–1154, 1987.

The Concrete Society, Diagnosis of deterioration in concrete structures. Technical Report 54, The Concrete Society, Crowthorne, 2000.

Woods, H., 1968, *Durability of Concrete Construction*, ACI Monograph No. 4, American Concrete Institute, Farmington Hills, MI.

3 Problems in Quality Management System

3.1 INTRODUCTION

As we discussed in Chapter 1, the construction phase is the main phase that causes a problem to the structure along its life time and the high percentage of structure failure due to construction error.

Therefore, the quality management system, in general, is the most critical management system line that we should follow strictly to achieve a safe structure. The total quality management (TQM) system should be followed from the start of the project from the design phase until commissioning and start up.

It is important to highlight that, in the mind of most civil engineers who is working on the site that there is a magic word "Factor of Safety" really this statement is famous due to lack of understanding the cause for a lot of problems or structure failures. This statement makes a lot of engineers on site they have imagine to be relaxed then the failure will happen. As we discussed in the previous chapter, the factor of safety in any code such as in British standard, BS, is the concrete partial safety factor and steel partial safety factor and the strength reduction factor in ACI code. The load factors also in the codes. All of these factors are not only to cover the uncertainty in concrete and steel geometric and physical properties but also to cover the variation in dead and live loads.

In this regard, we will focus on quality systems in construction projects and clarify the responsibility of all parties that share in the project and their responsibilities toward TQM through the management of the project.

All companies and organizations that have a target to grow their business in the open market now and a need to share in a suitable part of their market locally and internationally are going to apply the TQM concept. In this chapter, it is important to keep focus on the way to apply the TQM in construction projects for engineering, contractors, and any supplier of the materials in these projects.

As mentioned earlier, the oil and gas business' concentration on low cost only is not good, but we usually focus on the project time to achieve the quality of the work and the materials that will not affect operations in future.

3.2 QUALITY SYSTEM

Due to globalization, you can purchase any materials or utilize any contractor or engineering services to your project from any place worldwide, generate the need to have a system to let the client guarantee that the purchased materials will have the required quality, and deliver in the time agreed between the client and salesperson.

Therefore, the client needs a system that gives him or her confidence in the products being purchased, thus decreasing the level of the risk in his or her project.

On the other hand, all the companies now for owner contractors, engineering, and vendors are multinational companies. Therefore, the main office may be located in the U.S., and the other offices may be in the Middle East, Asia, or Europe. Therefore, can you imagine how controlling these offices give a guarantee that the offices that are far away will deliver the product to the client in good quality and will protect the company reputation. So many are searching for a third party who can guarantee that the product will be delivered in a good manner.

For a long time, there was no third party capable of providing this confidence that a factory can provide good quality materials. So, it becomes a real need to have specifications in order to achieve the quality assurance. Some specifications have been developed that control all steps of execution and manufacturing.

Work on these specifications began in the United Kingdom through the British Standards Institute, as it has been publishing a number of instructions on how to achieve the BS4891 quality assurance. After a short time period, a number of acceptable documents were made to meet the needs of the manufacturer or supplier.

From here, the specification BS5750 began and was published in 1979 through a series of specifications, which are guidance for internal quality management in the company as well as quality assurance of the product from outside the company. Quickly standards became acceptable to the manufacturer, the supplier, and the customer. Then the BS5750 standards became the benchmark for quality in the country.

At the same time, the U.S. was preparing a series of specifications, ANSI90, through the American National Standard Institute.

Some European countries also began preparing specifications in the same direction. The British specifications are considered to be the base point for any European specifications.

3.3 ISO 9001

It is common to hear about ISO and its relation to quality. The International Organization for Standardization (ISO) was established in 1947 as an agent for the United Nations, and it consists of representatives from 90 countries, sharing in the BSI and ANSI.

The activities of the ISO increase with time, and there are many specifications published through the ISO. The specifications have widely spread due to the interest from manufacturers, their international agents, and customers. The manufacturer provides a product to give the customer satisfaction, which increases the production and sales.

Therefore, now machines and equipment are designed and manufactured in conformity with the international standards to guarantee that they will be acceptable for use in all countries, which increases the volume of sales and marketing of products.

The ISO 9000 specifications were released in 1987 and were very close to British standards BS5750, parts 1, 2, and 3. The same general arrangement of the parts and the ISO increased as a general guide to illustrate the basic concepts and some applications that can be used in a series of ISO 9001.

On December 10, 1987, the board of the European committee for standardization agreed to work on the specifications of ISO 9001. Also, it is formally considered a standard specification for European countries without amendments or modifications, and it was published in the EN29000 1987. The official languages of the European standards are English, French, and German, and this group agreed to publish and translate these specifications for every country based on its language. Then the development of such standards was in 1994 when about 250 articles were modified. The articles often clarify the specifications and make it easy to read, and with time the number of countries working with these specifications increased.

The definition of quality management in ISO 9001 is an organizational structure of resources, activities, and responsibilities, which provides us with procedures and means that make us trust in the ability of the company and any institution, in general, to achieve the requirements of quality.

3.4 QUALITY MANUAL

A company's quality manual is the formal record of the firm's quality management system. It can contain the following:

- A rule book by which an organization functions
- A source of information from which the client drive confidence
- A vehicle for auditing, reviewing, and evaluating the company's QMS
- A firm statement of the company policy toward QC
- A quality assurance section and description of responsibilities

The source of information for the client should ensure the organization's ability to achieve quality. Identify responsibilities and relationships between individuals in the organization. This manual is the commander of the process, review, and evaluation of the quality system within the institution.

Therefore, this book must contain the following:

- Models of documents as well as models for the registration of the test results
- The necessary documents to determine how to follow-up on quality

3.4.1 QUALITY PLAN

The quality plan contains steps to achieve quality in a practical way with the order of action steps to reach the required quality in the project.

The quality plan varies from one project to another according to the requirements of the contract with the owner or the client. In general, the contract provided to achieve quality in a certain way and select a particular goal that needed a plan of action for quality in order to reach what was requested by the client. The quality plan should include the resources he will use, different types of personnel, and equipment to achieve quality. It should also identify the responsibilities, methods, procedures, and work instructions in detail, illustrated with a program of testing and examination.

It is worth mentioning that the quality plan must be inflexible and cannot be edited, making it stable with time until the end of the project.

Some contracts require that the buyer, client, or owner projects have special requirements that are needed to achieve the final product, which must be clarified in the quality plan, detailing what steps should be taken to achieve what the client requires. This plan should be presented to the client to trust in the ability of the plant to achieve the quality of the desired product. The plan should include the following:

- All controls, processes, inspection equipment, manpower sources, and skills that a company must have to achieve the required quality
- QC inspection and testing techniques that have been updated
- Any new measurement technique required to inspect the product
- No conflict between inspection and operation
- Standards of acceptability for all features and requirement that have been clearly recorded
- Compatibility of the design, manufacturing process, installation, inspection procedures, and applicable documentation have been assured well before production begins.

Quality control should involve company executives as well as field personnel. Quality control plans provide the written "reference" document for the implementation of the quality control program.

This plan must explain the duties and activities of the quality control personnel as clearly and concisely as possible. The following writing suggestions should be used due to drafting such a plan.

The plan should be from the different departments involved with the quality control process. Also, it includes the field office personnel and the participation of the owner, engineer, subcontractors, and suppliers.

Preparation and implementation of the QC plan must be more than a "cosmetic" fix. The quality control program may look good on paper, but it can only serve its intended purpose by daily execution of the stated quality control procedures.

The plan must be easily understood by the person who is going to implement the procedure listed in the manual. Items that should be included are organizational charts showing the chain of command, explanation of duties, lists of procedures, and examples of documents to be used.

The plan must be kept up-to-date by reflecting all changes required to maintain effective quality control on the job-site. This may include using suggestions from the employees responsible for QC duties.

The following guidelines have been established by the U.S. Army Corps of Engineers to be used when writing an organizational chart for the QC department:

- Lines of authority
- QC resources
- Adequately sized staff
- Qualifications of QC personnel
- List of QC personnel duties

Problems in Quality Management System

- Clearly defined duties, responsibilities, and authorities
- Deficiency identification, documentation, and correction
- Letter to QC personnel giving full authority to act on all quality issues
- Letter stating responsibilities and authorities addressed to each member of the QC staff
- Procedures for submittal management
- Submittals must be approved by the prime contractor before review by owner's representative
- Log of required submittals, listing all required submittals showing scheduled dates that submittals are needed
- Control testing plan
- Testing laboratories and qualifications identified
- Listing of all tests required as stated by contract documents
- Testing frequencies listed
- Reporting procedures
- Quality control reporting procedure addressed

3.4.2 Quality Control

The quality control definition in ISO is a group of operations, activities, or tests that should be done in a definite way to achieve the required quality for the final products.

In construction projects, the final product is the building or structures, which should function properly. So, in this case, the first step in quality control for getting the final product is to define the level of supervision in all the project phases to be sure that every part of the project is performed properly according to the required specifications. Try to ensure that the design, execution, and the use of the buildings and structures are compatible with the project specifications.

Note that the quality control is responsible from any level from the manager to lower levels in the organization. In a practical case, the construction managers and the department head are responsible for quality control, but it should be clear that everyone is responsible for quality control, except the sponsor.

3.4.3 Quality Control and Building Failure

The improvement of quality provides many benefits. The use of quality control will lead to fewer mistakes by ensuring that work is being performed correctly. By eliminating the need for corrective rework, there will be a reduced waste from the project resources. Lower costs, higher productivity, and increased worker morale will then lead to a better competitive position for the company.

For example, consider two crews. Assume that each of the crews has the same crew size, skill level, and work activity. However, the first crew takes the benefit of having another person perform quality control duties. Therefore, if any defective work is built, it can be corrected before work proceeds any further. Any defect in the work made by the second crew will probably be discovered after the work is completed. This defect in the work will be torn down and corrected or ignored and left in place. Then, the latter choice will cause problems as construction progresses and will provide the owner with a degree of dissatisfaction. Customer dissatisfaction can cause the company to be removed from consideration for future construction projects or could require a costly correction due to the amount of work affected by the correction.

Additionally, defects are not free. The person who makes the mistake has taken money and the person who corrects the defective work will obtain money, too. Additional material and equipment costs will also apply to this correction process.

One example showing the effect of defective work is seen in the partial collapse of a parking garage in New York City. The absence of reinforcing steel in three out of six of the cast-in-place column haunches, which supported the main precast girders, was the cause of this accident. The project plans and the rebar shop drawings showed that reinforcing steel was to be installed at these locations but was accidentally left out. As a result, extra work had to be performed at the contractor's expense to correct the work and repair the damage post.

Another major quality blunder occurred when constructing a shopping mall in Qatar. After pouring the concrete for the columns and the slab, they found around 40% of the columns had a strength lower than the allowable strength. So due to the lack of concrete quality control on site and experience of the staff, this cost a lot of money to repair and also delayed the whole project.

Quality is often "sacrificed" to save time and cut costs. However, quality does indeed save time and money. Nothing saves time and money more than doing the work the right way from the first time and eliminating the rework.

3.4.3.1 Submittal Data

The submittal data are usually the shop drawings, samples, performance data, and usually all the deliverable materials are required data test results or, Letters of Certification. The review of submittal data is one of the first steps in the quality control process. The information received from subcontractors and suppliers for items to be installed into the project must be verified to meet the standards set forth in the contract documents. Items such as dimensions (thickness, length, shape), ASTM standards, test reports, performance requirements, color, and coordination with other trades should be reviewed and verified carefully. Checking submittal information is important especially when shop drawings are checked. Due to the fact that the contract drawings do not provide enough detailed information to fabricate material, vendors and suppliers must make shop drawings. Materials that require shop drawings include concrete reinforcement, structural steel, cabinets/millwork, and elevators.

Basically, any item that is fabricated off-site is required to have a shop drawing in the submittal information. It is the information provided on the approved shop drawings that the fabricators use to "custom-make" their materials. Therefore, each item on the shop drawings must be verified against the contract plans and specifications. Once the submittal data are meticulously reviewed, whether it is a set of shop drawings or some other form of submittal data, a determination on whether or not the data should be "approved" or "disapproved" must be made.

Submittals are reviewed by the General Contractor and they are given over to the consultant engineer for further review. If the data are "disapproved" or incomplete, the originator of the submittal data must resubmit correct or additional information.

Problems in Quality Management System 31

A submittal (set of submitted information for a proposed material/equipment) that completes the review process is used as the "template" by which the material is fabricated. Any mistakes not discovered in the submittal review process will lead to potential problems involving extra cost and additional time for correction.

The Kansas City Hyatt Regency walkway collapse in 1981 is an example of how a poor shop drawing review can lead to disastrous consequences. In this case, a change in the details of structural connections, left unchecked during the submittal process, basically doubled the load on the fourth floor walkway connections.

This extra load on these connections led to the collapse of the fourth floor suspended walkway onto the second floor walkway and then onto the ground floor below. This disaster led to around 114 deaths and over 200 injuries.

3.4.3.2 Materials Tests

Once submittal information is checked against the contract requirements and approved, it is filed for future reference. Many companies file submittals in reference number.

Verifying that incoming material meets contract requirements is accomplished by using the data shown in the submittal information. The information found on the delivery tickets or the manufacturer's information provided with the shipment is compared to the information given in an approved submittal. If all information is correct, then the material can be approved for off-loading to the storage site.

Take care that any "unapproved" material that is allowed to be stored on-site has the possibility of being included in the construction process and leads to rework or other corrective action. Therefore, it is important that each item coming onto the site is verified to comply with the contract requirements. The concrete materials tests and the materials deficiency that cause building failure will be discussed in Chapter 4.

There are two types of concrete tests that are used to evaluate concrete on the jobsite: the slump test and the concrete cylinder, or cube test. The slump test, per ASTM C143, determines whether the desired workability of the concrete has been achieved without making the concrete too wet.

The project specifications for mortar shall mention that mortar must comply with either ASTM C270 or ASTM C780.

ASTM C270 states the required proportions of mortar ingredients (one part Type S masonry cement to three parts masonry sand), while ASTM C780 states the method of obtaining samples for compressive testing and the strength required for the mortar. Copies of these ASTM standards must be obtained to ensure full compliance with both the project specifications and industry standards.

3.4.3.3 Methods of Check the Construction Activity

The layout of work and the verification of correct placement, orientation, and elevation of work are extremely important. Work that is not placed correctly will lead to an extra cost for rework. For example, the misplacement of anchor bolts for the foundation will lead to expensive correction work and delays.

In addition to checking work, the proper layout of work is also required. The required tools needed to perform this function include the use of a tape measure, plumb bob, carpenter's level, and a chalk box. Topics to discuss for the proper layout and checking of work include checking elevations at the height of concrete footing during placement and finishing the grade and floor. Methods to check for proper alignment of work in the field manufacturer's recommendations for the layout of certain items are windows, overhead door, and air-handling units.

Since quality control is the responsibility of everyone involved in the construction process, most of the engineers in construction positions will help to manage QC functions. Since it is not always clear what one needs to find in order to ensure a proper inspection, engineers should be instructed to watch for "key items" during inspection.

As an example of QC Items for Steel Door and Frame Installation will be as follows:

1. When delivered to the site, each door and frame should be checked for damage.
2. Ensure proper size and gauge of doors.
3. Doors and frames must be stored off the ground in a place that protects them from the weather.
4. Do not stack doors or lay doors flat. This will cause doors to warp. Doors must be stacked on end of a carpet-covered racks or using other appropriate methods.
5. Check doors and frames for proper material, size, gauge, finish (satin, aluminum, milled), and anchorage requirements.
6. Verify door installation per door schedule shown in contract documents.
7. Fire-rated doors or frames must be used in fire-rated wall assemblies.
8. Fire-rated doors and frames must have a label attached or a certificate stating the fire-resistance rating.
9. Check for the proper location of the hinge side of the door and for proper swing of the door. (For example, per fire codes, the door swing for stairwells and other egress openings must open out, not into the stairwell.)
10. Door frames in masonry walls must be installed prior to starting masonry work (masonry must not be stepped back for future installation of door frame).
11. Is the doorframe installation straight and plumb?
12. If wood blocking is required for doorframe installation, make sure this activity is completed during the construction of the wall.
13. There is a uniform clearance between the door and doorframe (usually 1/8").
14. Has adequate clearance been provided between the bottom of the door and the floor finish (carpet, tile) that will be installed?
15. Touch-up scratches and rust spots with approved paint primer.
16. Exterior doors must be insulated.
17. Check for weather-stripping requirements on exterior doors.
18. The intersection between the doorframe and wall should be caulked—check for missing caulking in hard-to-reach areas (e.g., hinge-side of doorframe).

Problems in Quality Management System

3.4.3.4 Material/Equipment Compliance Tests

Every materials and equipment shall be tested prior to placement and after installation. Engineers and quality control team onsite should be familiar with testing methods, whether or not they will be performing the actual tests. Prior to beginning construction operations, a listing of each test that will be required should be made out.

This will serve as a checklist to be used by QC personnel. This testing checklist should list the type and frequency of testing required per each segment of work. Once tests have been performed, a test report documenting the results of the test should be kept on file or put into a "test report" folder for future reference. The following tests are typical tests that will be performed on the jobsite to ensure the quality of work is placed or completed.

3.4.3.5 Soils Testing

The foundation of a structure is responsible for transferring the loads from that structure into the ground below. The soil in this ground must be strong (dense) enough to stand with the loads that will be imposed. Additionally, the strength of soil must also be uniform to avoid any differential settlement in the structure, which can possibly cause structural and weatherproofing problems. In order to ensure that minimal settlement takes place in the building structure, the compaction of the soil must be verified. Each excavation or soil backfill operation must be checked to ensure compliance with the compaction requirements listed in the project specifications. These tests should be done before start any construction. Just before design the foundation.

As described in Chapter 1, a lot of problems and structure failure due to settlement and poor foundation design. This is due to the deficiency in soil test or the little confidence on the soil report data analysis and do not consider the recommendation in design or construction will cause a lot of problem and serious failure. It is very important to highlight that you should select a qualified geotechnical consultant office and sure that his office and lab are follow the quality management procedure.

3.4.3.6 Inspection During Work-in-Progress

In some cases, the inspection of work-in-progress must be performed on a continual basis. QC personnel must maintain constant watch on work as it begins and heads toward completion. It is very important to verify that work starts out correctly; otherwise, rework to correct the problem will occur. It is easier, and less expensive, to correct work as the work progresses instead of discovering defects after the work is completed. No one likes to perform the same item of work more than once.

3.4.3.7 Pre-installation Inspection Reports

Pre-installation inspection report forms are helpful due to trying to schedule an inspection for work-in-place prior to being covered up by the next phase of work. These forms are signed once the stated portion of work is completed. The general contractor's quality control personnel perform their final inspection once everyone else has "signed off" on their portion of work. However, it should be noted that quality control inspections have to be performed on a continual basis while work is being performed. These pre-work installation forms are used for final inspection purposes, not for the initial inspection of the work.

3.5 QUALITY ASSURANCE

The following is an example of the importance of quality assurance. You decide to build a new building, and about 7 years ago a contractor company built you the same building in high quality in the plan time and cost. Assume you are responsible for the decision without any influence from others.

Is it a good decision to go directly to this company or not? Why? (Please answer these questions before going to the next paragraphs.)

There are now different multinational companies worldwide in different industries, and one of these industries is construction. So, every company and every one of us are both a customer and manufacturer or service provider sometimes. For instance, the contractor company does the service to the client and, at the same time, this contractor company is a client of the manufacturer for the plumbing equipment, HVAC, ceramic tiles, and other materials and equipment required to complete the project. At the same time, the factory that sells the ceramic tiles is also a client to the mechanical spare parts company to maintain their machines working.

So, any defect to any one of the systems will affect all. It is obvious that the quality system should apply to all the companies and organizations, assuming that everyone in the company has a good quality system.

Every company should build its own system to ensure that the product and service is based on specifications, requirements, and satisfaction.

When the quality assurance system is strong, it means that if anyone in the organization moved or retired the quality of the product is the same.

As an answer to the first question, if this company is a family company with a father and a son, and the son becomes lazy and doesn't care, the project could be in trouble if his company is sharing in any of the project activity. On the other hand, if he has a real quality system, you can deal with him but you should also do an audit as we will explain later.

On the other side, for the multinational company the chairman is usually sitting in a country far away from the project, so quality assurance will be in a document that can be reviewed by an external or internal audit. If there are complains about the company from the owner, the system should record and solve these complains.

The purpose of quality assurance is as follows:

- To make sure that the final product is in conformity with the specifications, and the employment is highly qualified and able to achieve a high quality of the product through the administrative system
- To ensure the application of the company's fixed characteristics among all sectors in the factory, regardless of the presence of the same people
- The benefits of the application of quality assurance systems can be summed up in that it gives the ability to produce a product identical with the required specifications and also to reduce manufacturing cost because it will reduce

waste or defective products. In particular, projects have a major impact because in these projects the time factor is very important and may be the main driver of the project.

For example, due to the construction of hotels, the provision of any day of the total time for the project will have a significant return on the owner. The same is true for oil projects. Therefore, when reducing or not rejecting any product, time is not wasted in removing what had been done or repaired or in negotiations between the team of the contractor and the owner and the supervisory, achieving savings of the total time for the project.

If the product is proper, no part is rejected form it will be strong and good relationship is achieved between the seller and the client by reducing the number of complaints from the client and with primarily that relationship is very clear between the contractor and consultant or architect of the owner. When the Contractor provides the service or the work required of him with the presence of complains and a few observations and do not have a strong impact on the whole project from the supervising engineer indicating that the quality of the work of the contractor.

But in the case of repeated complaints from the contractor and that have a strong impact on the project, the problems may occur in the project, and the result is that he would not be called again in similar jobs and this is so dangerous in the world of markets as the reputation and quality of the final product with a good relationship between the parties has a powerful and direct influence in the reputation of the company, which provides the final product. The quality assurance system is the basic for any factory, construction company, or owner to have the ability to enter both internal and external competition.

So, the answer of the question is that we can use the same contractor, but we shall do a verification to its existing quality management system. Many of the problem onsite is due to the qualification of the contractor or the subcontractor he provides these can cause a serious problem during the project or after receiving it.

3.5.1 The Responsibility of the Contractor

The definition of contractor refers to the one who engages in the construction or supplying of the materials, as well as the one who is supplying the service required of it. In all references of study, control and quality assurance have been identified. In the case of construction projects, the company is the contractor for the establishment and implementation as well as the Engineering Office, which offers the service represented in the design and preliminary engineering drawings.

It is clear that the first responsibility during the process of TQM is the responsibility of the contractor. Whether that is the office engineering consultant or construction company, they must make sure the plant and anything or everything that comes out through this foundation must match the specifications required by the owner or the party of the owner, if the owner is represented in a company or institution.

The result became moving different products between countries and continents with free trade. Now there's talk that the world is smaller through international trade agreements, and its attendant laws and mechanisms help the trade between countries, which now has an impact on the industry with its various types. So it rests with you to achieve the requirements of the open market, which has led to fierce competition among different companies in the field of construction. We find the presence and proliferation of offices, international consulting, or multinational contracting companies exist on the map of competition in the Arab world. The competition between these companies stems primarily from the followers of quality assurance systems. In fact, the competition, conflict, and their interaction led to the presence of some different administrative regulations.

For comprehensive quality and emphasized quality control, work had been done to make these systems helpful for companies to deal between different countries. The overall quality assurance systems depend, basically, on customer satisfaction achieved by adequate revenues, which help them get a good reputation in world markets and allow them to compete.

The knowledge of environmental requirements of the state may affect the quality of implementation and the return of the project since the designer designs piles on the basis that they are close to the residential community and the work on these piles might exceed the permissible limits of noise. Thus, it requires a change in design such as the use piles of discharge.

There are some basic steps that should be important to the company, which deliver the service to improve quality. The main steps include that the senior management level would be interested in the importance of control and quality assurance through comprehensive quality leadership.

The second step is that the management level would provide an atmosphere that helps in dealing with the rules of quality assurance easily and make sure that all employees are following instructions and steps of quality assurance. These reasons are often constraints faced by the administrative level, and they are raised by engineers and junior staff.

Senior management should pay close attentions to the training process by organizing training courses for all the employees of an organization on the quality assurance procedures and technical labor in particular.

3.5.2 The Owner Responsibility

It is noted that many problems arise because of the bad quality of the final product or non-conformity of the project to the required specification, which is the fault of the owner or applicant because he or she may not have defined the desired product or specifications clearly. Therefore, the contractor must have all the required and completed data, and this is the responsibility of the Consultant Office of the owner.

Based on the specifications, the contractor shall determine the price and schedule based on the quality of the product itself, and it is the responsibility of the owner to identify the required specifications of the project strictly achieving its objectives. The selection of the contractor or the manufacturer is one of the most important and most serious responsibilities of the owner or representatives of the owner.

First, the owner selects the Engineering Office, and then they must choose a contractor, as these two selections are important factors and fundamental to the success of any project. Therefore, it is the responsibility of the owner or his representatives to gather enough information about the engineering office and the contractor and ensure from their previous work experience and that they had performed the same project before. In the vital project the owner can review the financial situation and make sure that the company is able to fulfill their obligations to the delivery of the project.

3.6 PROJECT QUALITY CONTROL IN VARIOUS STAGES

The project has been defined as a set of activities that has a beginning time and a time to be finished. These activities can be different from one project to another depending on the project. There are cultural projects or social projects, such as the literacy project, projects in engineering, such as the establishment of a residential building in an office or hospital or the construction of a full apartment, or industrial installations, which are called construction projects for roads, bridges, or railway. There are also irrigation projects and projects called civilian, but here we will focus on construction projects and specifically oil and gas projects. Construction projects vary from one project to another depending on the size and value of the project. Therefore, the degree of quality control varies depending on the size of the project, especially in developing countries. Quality control may be sufficient in small businesses, but the contracting companies or small engineering offices that are aimed at the international competition are also increasing the quality of the projects, which increases the total cost of the project.

It is important to control the project during all steps in the life cycle the main stages that can cause a structure failure is the detailed engineering phase and construction phase.

3.6.1 DETAILED ENGINEERING

It is assumed after the end of the preliminary engineering stage that one can get the complete drawings and the full specifications for the whole project, containing all the details that make the contractor able to carry out the works. Those drawings are called the Construction Drawings.

Therefore, this phase requires many working hours, extensive contacts, good coordination, and excellent organization. A good manager allows freedom of communication between individuals and also allows review with strong and continuous coordination.

It is noted that the complexity of this stage needs an affirmation of the quality.

Engineers always believe in utopias where our lives are dependent on the accuracy of the accounts and reviews, but the teamwork is not always so accurate. Often work reaches someone late, or you may have to take actions to correct it, or there may be some change in procedures within the company or department without your knowledge. All of this leads to a loss of time, and therefore we believe that the overall atmosphere in which we work needs to be reformed, and this itself is a vision system of quality assurance.

The system of quality assurance provides stable functioning of all departments; however, there may be a change in personnel. This problem often occurs at the stage of studies where you need it for intensive cooperation.

For example, when there is a strong relationship between the managers of the departments of civil and mechanical, you will see a sharing in information, the work will run smoothly, and there will be periodic meetings and productive correspondence.

The system of quality assurance at this stage is important because it is a process of organizational work. Everyone should know the goal of the foundation and the goal of the project by the institution, and the responsibility of each individual and the concept of quality is clear at all times and supported by documentation.

Documents are considered the operational arm of the quality application process and must, therefore, be in the event of any amendment or correction in the drawings. When you set up the drawings, they must be received at a specific time for the owner to review, and discussion takes place with any amendment, and remarks are to identify changes.

Through the development of a particular activity, it may be canceled, so there should be a quality system procedure in place to avoid confusion with the other copies/versions of drawings and eliminate the human error. The modification revision number will be continued updated and this system will be continued until we reach the final stage of the project and take the final approval of the drawing with sealed stamp, indicating that it is the final drawings for approval of construction. "Approved for Construction" drawings are obtained after the completion of the study phase and start the execution phase should have specifications and drawings fully ready to start the construction phase. You can imagine that in some of the projects may reach the hundreds of drawings and have your special specifications and other operation folders manual, as well as for maintenance and repair in the event of some failures or trouble shooting.

Generally, it can be summarized in the following principles of design, which are divided into five aspects and should be covered in each quality assurance system:

1. Planning, design, and development—Determine who does what in the design.
2. Entrance design—Be sure you know what the client wants in the design.
3. Troubleshooting design—Provide clarity for the final form of the design.
4. Verification of the design—Review with the client to make sure that the design is consistent with the needs of the client.
5. Change design—Ensure that any change in design will be adopted by responsible

3.6.2 Execution Phase

Now everything is prepared for this stage, and this stage requires both quality assurance and quality control. In the work of reinforced concrete structure, concrete itself consists of various materials such as cement, sand, gravel, water and additives in addition to steel reinforcement and, therefore, must be controlled in terms of quality of each article separately as well as in the same mixture. The above must be done for quality control during the preparation of wooden forms and assembling the steel, casting, and processing.

Thus, the contractor should have the administrative organization well organized in order to achieve quality control in addition to the existence of documents that

Problems in Quality Management System

would identify the time and date when work is carried out to determine the number of samples of concrete, which is to test pressure resistance and to identify the exact time, date, and test result.

Often during construction there are some changes in the drawings as a result of the emergence of some problems at the site during the implementation or appearance of some ideas and suggestions that reduce the project time. Some changes do not negate the imbalance in the quality assurance system in the design stage.

After the work, the change must be made in the documents, and post implementation is modified to get the drawings identical to the site of the as-built drawing.

The supervisory authority and the owner must both have their own organization, and the two most common cases in projects are the following:

1. Owner has the supervision team on site.
2. Owner chooses a consulting office that performs the design and handles the supervision.
3. In both cases, there should be strong organization, similar to the organization of the contractor, as it is the tools that control quality. But when the contractor has a working group who has full knowledge of quality assurance and control, the dispute between the supervisory and the contractor will be more narrower, where the controversy will not be about the final quality of the project. It is very important to the owner that the project is in good shape in the end.

You can imagine that if the owner has a competent staff and a strong knowledge about QC and QA while the contractor staff has no any knowledge about it, there will be more trouble that will affect the project as a whole.

The construction phase shows the strength of the contractor, its international competitiveness, and local communities if the concept of quality assurance for the work team is quite clear, because all the competitors on the international scene for some time have been working through an integrated system aimed at assuring the quality of work and set quality in all stages of implementation in order to achieve total customer satisfaction.

3.6.2.1 Inspection Procedures

The phase of execution will be supplied by all required materials from many different locations in addition to the installation of these materials, and it is the responsibility of the general contractor to make sure that the supply of various materials as well as the construction and installation must be done through the required quality. In the case of construction and installation, there must be specific instructions to determine the manner in which they work and the proper equipment to be used. Investigation is the final quality process of the work or the quality of the materials supplied. It requires continuous inspection and testing work, and the inspection must specify the following:

- The substance to be tested
- Test procedure

- Equipment required for testing and inspection and calibration of such equipment
- Inspection method
- Environmental conditions required, which will maintain it during operation, inspection, and testing
- The sampling method or the way to choose the appropriate sample
- Defining the limits acceptance and rejection of the samples tested

The following items are from the ISO 9001 Section 4.11 and the inspection and measurement test.

- Control inspection
- Measurement and test equipment
- Calibration, maintenance, and the surrounding environment, and storage and documents
- Registration and inspections

3.6.2.2 Checklists

The ISO 9001 selects who conducts the review process and some of the menus that contain questions from the manufacturer or internal departments. When you read the questions in the lists, you will find that they cover many important aspects of basics and quality control in all stages of the manufacturing or implementation of product.

Checklists contain the following specific questions, which are with the auditors of public review of the company's performance with other lists in the special design phase and implementation phase, and those lists are detailed with questions fully to control the full and comprehensive review of the quality system at the stage of design and implementation.

3.7 EXTERNAL AUDITING FOR THE PROJECT

It is important to highlight that in major projects you shall take care that may be some companies have the ISO certificate but there are not follow the procedure, so even if they present the certificate to you should perform auditing to their facilities.

The auditing shall be done by a team from your side from the quality manger and one from the discipline engineer that will receive their product or the service.

To be sure that the company will be welling for this visit from the owner auditing, he will gain a lot from the big project but it will be difficult in small projects. So, to control this issue a company needs to be registered with your company, and they should provide their prequalification. At this time you can perform the auditing to decide if they will be registered with companies' vendor's lists or not.

The team from the owner company should be competent and have strong skills and knowledge in quality and how to audit the other company in the country or overseas countries.

Problems in Quality Management System

First of all, the contractor or the service provider will deliver his quality manual, which will be reviewed by the quality team. Then they will visit the site with a representative from the company to show their system in action and to inspect it. The process is as follows:

- Visit the supplier site to perform complete inspection.
- The supervisor will describe to the team exactly how their QC system works.
- The contractor provides examples of QC documentation.
- It is possible for the team to ask for a previously inspected batch to be rechecked.
- Check if the test equipment is regularly maintained.
- The rejected or unacceptable products are clearly marked and segregated to avoid any chance of their accidental inclusion with acceptable products.

After the site visit, there are three probable outcomes:

1. Acceptable to be registered in company bidder list

    If the evaluation has shown that the supplier has a satisfactory QMS, there are no deficiencies and the supplier is able to give an assurance of quality.

    Also, in this case the supplier may have proven that they are up to satisfactory standards.

2. Weak Quality System

    If the team finds several significant weaknesses in the supplier's system, the supplier will have to take steps to overcome these failures and improve their QMS.

    The supplier can ask for another evaluation to confirm that their quality is approved. To enhance their company and employee work will take time, so it is not preferred to register the company but to give it a period to change.

3. Unacceptable Quality System

    In this case the team found that the supplier would have to make radical changes to improve their overall QMS.

    Note that in this case the actions from the supplier to reach the target to satisfy QMS will take not less than one year, so avoid dealing with this supplier.

ISO 9000 provides a check list that is essential and important for the auditing team. The following is the check list for every phase of the project. In the design phase there is a sample for what will be asked from the engineering company as in Table 3.1. For the construction phase, this check list is tailored to be a sample when auditing a contractor company as shown in Table 3.2.

From the above it is important that all employees in the project have very good technical skills plus the quality system, and this is stated in ISO 9001 Section 4.18.

TABLE 3.1
Design Check List

Item	Questions	Yes/No	Remarks
1	Do they have a system to assure the client presents his or her needs clearly?		
2	Are the client requirements clear to all the design members?		
3	What is the international standard and specification they use?		
4	Are these standards available in their office?		
5	Are the drawings and documents sent by the client registered?		
6	Do they have a document management system?		
7	Do they define the name of the discipline lead?		
8	Are the activities clear to them?		
9	How can they select the new engineers?		
10	Do the drawings have a number?		
11	Are there strong numbering systems to the drawings?		
12	Do they prepare a list of the drawings?		
13	Are they updating this list?		
14	What is the checking system in calculation?		
15	What is the checking system in the drawings?		
16	Are the employees familiar with CAD?		
17	Do they have a backup system to the documents and drawings?		
18	Do they have antivirus?		
19	Do they use a sub engineering office?		
20	What is the method and criteria for the selection?		
21	Is there a good relation between the design team and the supervision team on site?		
22	Are they experienced with the technical inquiry?		

The should be required in training to increase their technical and managerial skills for quality control in addition to all the knowledge about TQM.

3.8 OPERATIONAL AND MAINTENANCE PHASE OF THE PROJECT

As we discussed before, about 25% of the building collapse occurred due to shortage in maintenance and operations methods and policy. The owner is responsible for the operation phase, as in this phase the owner will have full authority and responsibility to operate the project after the commissioning and start-up phase.

During operation, there is usually a requirement to modify the facilities due to operation needs.

In some regular cases, workshops are extended in normal industry. Most international companies that follow ISO have a management of change (MOC) system and procedure.

TABLE 3.2
Construction Checklist

Item	Questions	Yes/No	Remarks
1	Is there a quality control procedure?	1	
2	IS the QC procedure understood by all members and how?	2	
3	Does the QC match with the task?	3	
4	What is the way to assure with the tests?	4	
5	Is there equipment to do the test calibrated?	5	
6	How they do the test piping, concrete, welding, and what is the confidence for this test?	6	
7	Is a third party used for the test?	7	
8	What are the criteria for choosing the third party?	8	
9	Do they use a subcontractor?	9	
10	What are the criteria for choosing these subcontractors?	10	
11	Do they regularly maintain their equipment?	11	
12	Do they have a certificate for the cranes and wire?	12	
13	Is there a team on site that knows if the project is time or cost driven?	13	
14	Does their team know the project objective and target?	14	
15	Do they control the documents and drawings on site?	15	
16	Do they have a document management system?	16	

In this procedure, the required modification is identified and approval is granted from all the engineering disciplines. In new projects, it is preferred to go to the original engineering office that performed the design to do the engineering for this modification.

An example of bad MOC in an international hotel is the conversion of one room from normal use to be a planet land for entertainment. They put clay on the floor without performing any MOC process, and due to this heavy load, the floor collapsed and damaged four cars in the garage underneath the room.

From previous experience, projects in every step of engineering and construction concentrate on these goals, and usually the input from the operation is very little.

However, the operation is the end user but usually a limited number of operation engineers share in the project for many reasons, as the operation cannot release many engineers during the project phase to cooperate with the project team. On the other hand, the operation members who are usually one or two engineers just define what they need. In this circumstance, there is usually an expectation for not full satisfaction from the operations department.

BIBLIOGRAPHY

ACI 228.1R89, In-place methods or determination of strength o concrete, ACI manual of concrete practice, part 2: construction practices and inspection pavements, 25 pp, Detroit, MI, 1994.

ASTM C188-84, Density of hydraulic cement.
ASTM C114-88, Chemical analysis of hydraulic cement.
ASTM C183-88, Sampling and amount of testing of hydraulic cement.
ASTM C349-82, Compressive strength of Hydraulic cement Mortars.
ASTM C670-84, Testing of building materials.
ASTM C142-78, Test method for clay lumps and friable particles in aggregate.
ASTM D1888-78, Standard test method for particulate and dissolved matter, solids or residue in water.
ASTM D512-85, Standard test method for chloride ion in water.
ASTM D516-82, Standard test method for sulfate ion in water.
Ajdukiewicz, A.B., and Kliszczewicz, A.T, Utilization of recycled aggregates in HS/HPC, *5th International Symposium on Utilization of High strength/High Performance Concrete*, Sandefjord, Norway, June 1999, Vol. 2, pp. 973–980, 1999.
Ajdukiewicz, A.B., and Kliszczewicz, A.T., Properties and usability of HPC with recycled aggregates. *Proceeding, PCI/FHWA/FIP International Symposium on High Performance Concrete*, Sep., 2000, Orlando, Precast/Prestressed Concrete Institute, Chicago 2000, pp. 89–98, 2000.
Ajdukiewicz, A.B., and Kliszczewicz, A.T., Behavior of RC beams from recycled aggregate concrete, *ACI 5th International Conference*, Cancun, Mexico, 2002.
BS 882, 1992.
BS 812 Part 103-19, Sampling and testing of mineral aggregate sands and fillers.
BS 410-1:1986, Spec. for test sieves of metal wire cloth, 2000.
Bs EN933-1 1997, Tests for geometrical properties of aggregates, determination of particle size distribution. Sieving method.
Di Niro, G., Dolara, E., and Cairns, R., The use of recycled aggregate concrete for structural purposes in prefabrication. *Proceedings, 13th FIP Congress "Challenges for CONCRETE IN THE Next Millennium"*, Amsterdam, June 1998; Balkema, Rotterdam-Brookfield, 1998, V.2, pp. 547–550, 1998.
Egyptian Standard Specification, 1947-1991, method of taking cement sample.
Egyptian Standard Specification, 2421-1993 natural and mechanical properties for cement. Part 2: define cement finening by sieve No.170.
Egyptian Standard Specification, 2421-1993 natural and mechanical properties for cement. Part 2: define cement finening by using blain apparatus.
Egyptian Standard Specification, 2421-1993 natural and mechanical properties for cement. Part 1: define cement setting time.
Egyptian Standard Specification, 1450-1979 Portland cement with fines 4100.
Egyptian Standard Specification, 1109-1971 concrete aggregate from natural resources.
Egyptian Standard Specification, 262-1999 steel reinforcement bars.
Egyptian Standard Specification, 76-1989 tension tests for metal.
Egyptian code for design and execute concrete structures: part3 laboratory test for concrete materials, ECP203, 2003.
El-Arian, A., El-hakim, F., and Abd El-Aziz, M.I., "Effect of storage condition of ordinary Portland cement on its properties", M.Sc., Structural Department, Cairo University, 1985.
Elreedy, M.A., "New project management approach for offshore facilities rehabilitation projects", SPE-160794, *Adipec Abu Dhabi Conference proceeding*, UAE, 2012.
ISO 6274-1982, Sieve analysis of aggregate.
Kasai, Y., (ed.), Demolition and reuse of concrete and masonry reuse of demolition waste, *Proceedings, 2nd International Symposium RILEM*, Building Research Institute and Nihon University, Tokyo, Nov., 1988 Chapman and Hall, London–New York, pp. 774, 1988.

Mukai T., and Kikuchi, M., Properties of reinforced concrete beams containing recycled aggregate. *Proceedings, 2nd International Symposium RILEM "Demolition and Reuse of Concrete and Masonry"*, Tokyo, Nov., Chapman and Hall, London–New York, Vol. 2, pp. 670–679, 1988.

Murphy, W.E., discussion on paper by Malhotra, V.M., 1977. Contract strength requirements-core versus in situ evaluation. *ACI Journal, Proceedings*, Vol. 74, No. 10, pp. 523–5.

Neville, A.M., Properties of concrete, PITMAN, 1983.

Plowman, J.M., Smith, W.F., and Sheriff, T., 1974. Cores, cubes, and the specified strength of concrete, *The Structural Engineer*, Vol. 52, No. 11, pp. 421–6.

Salem, R.M., and Burdette, E.G., 1998. Role of chemical and mineral admixtures on physical properties and frost—resistance of recycled aggregate concrete. *ACI Materials Journal*, Vol. 95, No. 5, pp. 558–563

"The building commissioning guide," U.S. General Services Administration Public Buildings Service Office of the Chief Architect, April 2005.

Tricker, R., "ISO 9000 for small business," Butterworth-Heinemann, 1997. Project Management Institute Standards Committee. A Guide to the Project Management Body of Knowledge. Upper Darby, PA: Project Management Institute, 2004, 1997.

Van Acker, A., Recycling of concrete at a precast concrete plant. FIP Notes, Part 1: No 2, 1997, pp. 3–6; Part 2: No. 4, 1997, pp.4–6, 1997.

Yuan, R.L., et al., 1991. Evaluation of core strength in high-strength concrete, *Concrete International*, Vol. 13, No. 5, pp. 30–4.

4 Bad Materials Cause Failures

4.1 INTRODUCTION

There are many problems that will occur in a building and these will be due to bad selection of the materials itself or lack of doing the required inspection and tests.

Concrete consists of two main parts: the first part is the aggregate (coarse and fine) and the other part is the adhesive between cement and water. The main components are the coarse and fine aggregate, cement, water, and additives. These additives are added to the mixture to improve the properties of concrete for better durability. These additives must be within some definite percentage, as they create a poor impact if the mixture percentage is not accurate.

To obtain a concrete that conforms to the specifications, we must adjust the quality of each component in the concrete mixture. Therefore, in this chapter, the main necessary tests to cement, aggregate, water, and additives will be illustrated to obtain a high-quality concrete and to match the international standard and project specifications.

It is worth to mention that the concrete itself is weak under tensile stress, so we use steel reinforcement in concrete to withstand the tensile stress, which will increase the efficiency of concrete.

Nowadays, the reinforced concrete is an important element in the construction industry worldwide because it is cheaper in most cases than other alternatives and because of its ease of formation in the early stages, which allows us different architectural forms.

Steel is the most important element in reinforced concrete structures as it carries a larger part of the stresses. For this situation, the production of the steel bars must be under a strict system of quality control to make sure that they follow the international standards and project specifications.

These tests should be performed in an appropriate way with valid calibration devices to obtain the accurate results. The laboratory as a whole system should follow a quality assurance system, and the devices should be calibrated periodically. Moreover, the samples must be taken correctly according to standard specifications and all the necessary tests performed and reviewed based on the quality system.

4.2 CONCRETE MATERIALS TEST

In the following sections, we will describe the important field and laboratory tests for concrete materials as most civil engineers on site focus on the compressive strength and slump test only and don't pay attention to the other tests.

Note that the other test is important for concrete durability and its performance with the time. Moreover, for cube or cylinder compressive strength, it gives us result 28 days after concrete has become hard so any modification or rebuilding is very difficult. Materials tests are important to ensure that the concrete strength will achieve the project specification target.

4.2.1 Cement

Nowadays, there are different types of cement to choose from the different alternatives depending on the designer recommendation in the project specifications and the drawings based on the surrounding environmental condition and the required strength.

The main types of cement in British description and ASTM are shown in Table 4.1.

To ensure the cement quality, there are many tests to define the quality of the cement manufacturing or ensure the cement onsite is a match with the specifications. There are some important tests that applied in many international standards and define the limits of refuse and accept the cement as stated in British and American association of testing materials (ASTM).

There are many ways to take a sample. Based on the procurement agreement, losse cement or bags on site may simple.

It is very important to define the location and time to take the samples in the agreement between the supplier and the customer. There should be enough time allotted to take the sample to perform the required tests, obtain the results, and define the suitability of the cement to use.

The tests should be conducted 28 days from the date of receiving the samples.

When taking the samples, they must be in conformity with the specifications agreed upon. The general requirement of the samples in the standard specification is that the weight of one sample is not less than 5 kg and that the sampling tools are dry and clean. The sample may be taken individually via a digital sampling or via a number of samples at intervals spaced called composite sample. In the case of bulk cement, it is necessary to avoid upper cement layer by about 150 mm from the top.

TABLE 4.1
Cement Types

ASTM Description	British Description
Type I	Ordinary Portland cement
Type III	Rapid hardening Portland
	Extra rapid hardening Portland
	Ultra-high early strength Portland
Type IV	Low heat Portland
Type II	Modified cement
Type V	Sulfate resisting Portland
Type IS	Portland blast-furnace
	White Portland
Type IP &P	Portland-pozzolana
Type S	Slag cement

Bad Materials Cause Failures

Most sites receive the cement in sacs so the tests will be done by select random sacs and the number of sacs will be as follows:

$$\text{The number of samples} \geq (n)^{0.333}$$

It is important to highlight that if you are using ready mix concrete, they will do these tests as their target to use the minimum amount of cement and provide the required strength. Therefore, it is your responsibility as the consultant or owner that whether they are doing these tests as per the standards to review their quality assurance procedure as we discussed before. But if you are doing the mix onsite, then you are responsible to perform these tests. It is important to highlight the quality control it is not just a slump test or the cube or cylinder test after 28 days. All the below tests or any other tests are part of the quality control to have a durable structure along its life time.

4.2.1.1 Cement Test by The Sieve No. 170

The fines of cement are important factors that affect the quality of concrete industry in general. If the cement particle is big, it cannot completely react with water as there will be a remaining core in the cement particle the water will not reach it. The water propagates through the cement particles and starts to dehydrate, causing an increase in temperature, which is the main reason for forming the hair cracks, causing no stabilizing for cement volume.

As a result of that, in the case of an increase in the cement particle size that reduces the strength to the same cement content, increasing the fines of cement will improve the workability, cohesion, and durability with time and will decrease the water moving upward to the concrete surface.

Figure 4.1 presents the relation between the concrete strength and the concrete fines at different ages.

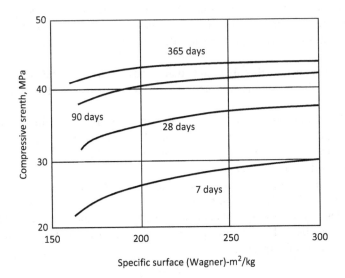

FIGURE 4.1 Relation between cement fines and compressive strength.

To do this test, take a sample of 50 g of cement and shake it in closed glass bottle for 2 min and then revolve the sample gently by using dry bar.

The sample will put in a closed bottle and left for 2 min. Put the sample in Sieve No. 170 (90 μm) and then shake the sieve horizontally and rotationally. Thus, the sieve test is finished when the rate of passing cement particles is not more than 0.5 g/min during the sieve. Now remove the fines from the bottom of the sieve by a smooth brush carefully. Then, collect and determine the weight of the remaining particles on the sieve (W_1).

Repeat the same test again with another sample. Then residual weight for the second test is obtained (W_2). The calculation of the sieve of the remaining samples is as follows:

$$R_1 = \left(\frac{W_1}{50}\right) \times 100$$

$$R_2 = \left(\frac{W_2}{50}\right) \times 100$$

The ratio is calculated (R) by taking the average of R_1 and R_2 to nearest 0.1%. When the results of the two samples differ by more than 1%, perform the test again for third time and calculate the average value of the three results.

Therefore, you can accept or refuse the cement based on the following condition:

- In the case of Portland cement, the remainder must not exceed 10%
- In the case of rapid hardening Portland cement, the remainder must not exceed 5%.

4.2.1.2 Define Cement Fines by Using Blaine Apparatus

This test is used to determine the surface area by comparing the test sample with the specific reference. The increasing of the surface area will speed of concrete hardening and obtains early strength. This test defines the acceptance and rejection of the cement.

There are many tests that are used to define the cement fines, and one of these tests is based on Blaine apparatus, as stated in many codes such as the Egyptian code.

- This test depends on calculating the surface area by comparing the sample test and the reference sample using the Blaine apparatus to determine the time required to pass a definite quantity of air inside cement layer with defined dimensions and porosity.
- In Blaine apparatus, as shown in Figure 4.2, first the total volume of the cement layer is determined by filling it with mercury and using two Blaine circles. Then the weight of cement before and after adding the cement layer is determined. Thus, knowing the mercury density allows us to calculate the volume of cement layer:

$$V = \frac{W_1 - W_2}{D_m}$$

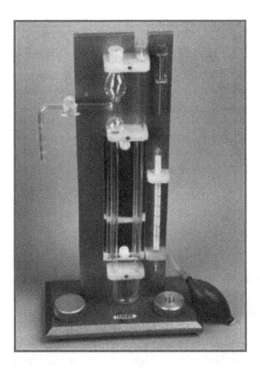

FIGURE 4.2 Blaine apparatus.

where V is the volume of cement layer (cm³), W_1 is the weight of mercury in gram (i.e., fills the device to nearest 0.0 g), W_2 is the weight of mercury in gram (i.e., fills the device to level of cement to nearest 0.0 g), and D_m is the density of the mercury (g/cm³). Now, use tables to define the mercury density at the average temperature of the test and by using the manometer in Blaine apparatus:

$$S_r = \frac{K}{D_r}\left(\frac{\sqrt{(P_r)^3 T_r}}{(1-P_r)\sqrt{0.1 I_r}}\right)$$

where S_r is the reference cement surface area (cm²/g), D_r is the reference cement density (g/cm³), P_r is the porosity of the cement layer, I_r is the air viscosity in the average temperature for reference cement test, T_r is the average time required to the manometer liquid to settle in two marks to nearest (0.2 s.), and K is the Blaine apparatus constant factor defined by the previous equation by knowing the time need to pass the air in the sample.

Thus, calculate sample surface area by using the following equation.

$$S_c = S_r \left(\frac{D_r}{D_c}\right) * \left(\frac{T_c}{T_r}\right)^{0.5}$$

According the Egyptian code which is a match with BS, the acceptable and refusal of cement will be as shown in Table 4.2.

TABLE 4.2
Cement Fines Acceptable and Refusal Limits

Cement Types	Cement Fines Not Less Than cm^2/gm
Ordinary Portland cement	2,750
Rapid hardening Portland cement	3,500
Sulfate resistance Portland cement	2,800
Low heat Portland cement	2,800
White Portland cement	2,700
Mixing sand Portland cement	3,000
4100 fines	4,100
Slag Portland cement	2,500

The question is who is doing this test and how can it cause a concrete failure? This test will be done by the ready-mix concrete laboratory as they will receive the materials; however, if the mixing will be done on site, one should do the test on site as well. The concrete design mixture is based on the cement standard with certain particle size. So, after mixing, if the cement particle is bigger the dehydration of the cement will not react with the whole particle. If the particle size increase doubles, the cement reacts about 50% of the design capacity and the strength will be less after 28 days.

4.2.1.3 Initial and Final Setting Times of Cement Paste Using VICAT's Apparatus

If you pour the concrete but it takes a lot of time until reach its final setting, as some structure member it will carry load after removing the shuttering in this case it will carry a full dead load. The objective of this test is to define the time for initial and final setting time to the paste of water and cement with standard consistence by using VICAT apparatus, as from this test can define if the cement is expired or can still be used.

It is known that the initial setting time is the time required for concrete to set and after that it cannot be poured or formed and the final setting time is the time required for the concrete to be hardened.

Vicat apparatus as shown in Figure 4.3 consists of a carrier with a needle acting under a prescribed weight. The parts move vertically without friction, weigh 300 ± 1 g, and are not subject to erosion or corrosion. The paste mold is made from a metal or hard rubber, or plastic like a cut cone with a depth of minus 40 ± 2 mm, and the internal diameter of the upper face is 70 ± 5 mm and the lower face is 80 ± 5 mm and provides a template of glass or any similar materials in the softer surface and porous. Its dimensions are greater than the dimensions of the mold.

The needle is used to determine the initial setting time in the form of steel cylinder with effective length 50 ± 1 mm and diameter 1.13 ± 0.5 mm. The needle measures the time of final setting steel in the form of a cylinder with length of 30 ± 1 mm

Bad Materials Cause Failures

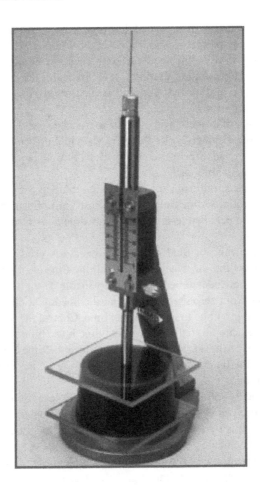

FIGURE 4.3 Vicat apparatus.

and diameter 1.13 ± 0.5 mm and installed by a 5 mm diameter ring at the party needle free end to be the distance between the end of the needle and the ring is 0.5 mm.

The test starts by taking a sample weighing about 400 g and placing it on an unpermeable surface. Then add 100 mL of water and record zero measurement from the time of adding water to the cement and then the process of mixing in a 240 ± 5 s on the unpermeable surface.

To determine the initial setting time put the vicat needle and calibrate the device until it reaches the needle the base of the mold. Then adjust the measuring device to zero and then return needle to its original place.

Fill the mold with cement paste with standard consistency and troll the surface and then put the mold for a small period of time in a place that has the temperature and humidity required for the test.

Transfer the mold to the apparatus under the needle, and then make the needle slowly approach the surface until it touches the paste's surface. Stop it in place for a

1 or 2 s to avoid impact of primary speed, leaving the moving parts to implement the needle vertically in the paste.

Grading depends on when the needle stops to penetrate or after 30 s, whichever is earlier, and record the reading of grading, which indicates the distance between the mold base and the end of the needle, as well as the time starting from the zero-level measurement.

Repeat the process of immersing the needle to the same paste in different locations with distance between immersing and edge of the mold or between two immersing points not less 10 mm and after consequent periods of time (about 10 min) and clean the needle immediately after each test.

Record time is measured from zero penetration to up to 5 ± 1 mm from the base of the mold as the initial setting time to the nearest 5 min. Ensuring the accuracy of measurement of time between tests reduces embedment tests and the fluctuation of these successive tests.

To determine the time of final setting, the needle is used to identify the final time of setting, follow the same steps in determining time of initial setting time and increase the period between embedment tests to 30 min.

Record the time from zero measurement until embedment of the needle to a distance of 0.5 mm and it will be the final setting time. Control the impact of the needle on the surface of the sample so the final setting time is the time presents the effect of the needle. To enhance the accuracy of the test reduce the time between embedment tests and examine the fluctuation of these successive tests. Record the final setting time to nearest 5 mm.

According to the Egyptian specifications, the initial setting time must not be less than 45 min for all types of cement except the low heat cement. The initial setting time should not be less than 60 min. The final setting time must be not more than 10 h for all types of cement.

If we ignore this test, it will result in a bigger problem as removing the shuttering based on the final setting and the time to gain the strength if the concrete mix will be set to fast can be a problem during construction. Also, they will cover the problem by adding water which will cause a significant decrease in concrete strength.

4.2.1.4 Compressive Strength of Cement Mortars

The cement mortar compressive strength test is performed with standard cubes of cement mortar. The cement mortar is mixed manually and compacted mechanically by a standard vibrating machine. This test is considered the refusal or acceptance test.

The compressive strength is one of the most important properties of concrete. The concrete gain its compressive strength from the presence of cement paste resulting from the interaction that occurs between the cement and water added to the mix. Therefore, this test makes sure that the cement used is to appropriate compressive strength to accept or reject it. This test should be done to all types of cement.

The equipment needs for the test are sieves with standard square holes of the fabric of wires opened 850 μm, 650 μm. Nylon made of stainless steel does not react with cement and weighs about 210 g.

Bad Materials Cause Failures

The vibrating machine has a weight of about 29 kg and the speed of vibration is about 12000 vertical vibration + 400 rpm and the moment of vibrating column is 0.016 Nm.

The mold of the test is a cube 70.7 ± 1 mm and the surface area for each surface is 500 mm^2. The acceptable tolerance in leveling is about 0.03 mm and the tolerance between paralleling for each face is about 0.06 mm.

The mold is manufacturing from materials that will not react with the cement mortar, and the base of the mold from steel that can prevent leak of the mortar or water from the mold, and the base should match with the vibrating machine.

The sand that will be used in the test must have a percentage of silica not less than 90% by weight, and must be washed and dried very well. Moreover, the humidity is not more than 0.1% by weight and this sand can pass through sieve opening 850 µm and the passing through the standard sieve size of 600 µm should not more than 10% by weight (Tables 4.3 and 4.4).

After performing the tests, the standard cubes will crush at age 1 day which is about 24 ± 0.5 h, and after 3 days within 72 ± 1 h, and after 7 days within 168 ± 1 h, and after 28 days within 672 ± 1 h.

Table 4.5 illustrates the limits of acceptance and rejection based on the cement mortar compressive strength. Note from the table that there is more than one type of high-alumina cement, high-alumina cements vary according to the percentage of oxide alumina.

Note that the compressive strength after 28 days will not be considered for acceptance or rejection unless stated in the deal between the supplier and the client.

TABLE 4.3
One Cube Mixing Ratio

Cement Type	Materials	Ratios by Weight	Weight (g)
All types of cement	Cement	1.0	185 ± 1
	Sand	3.0	555 ± 1
	Water	0.4	74 ± 1
High-alumina cement	Cement	1.0	190 ± 1
	Sand	3.0	570 ± 1
	Water	0.4	76 ± 1

TABLE 4.4
Allowable Deviation in Temperature and Humidity

Location	Temperature (°C)	Min. Relative Humidity (%)
Mixing room	20 ± 2	65
Curing room		90
Water curing sink		–
Compression machine room		50

TABLE 4.5
Acceptable and Refusal Limits

	Cube Compressive Strength (N/mm²)			
Cement Type	After 24 h ≥	After 3 days ≥	After 7 days ≥	After 28 days ≥
Ordinary Portland cement	–	18	27	36
Rapid hardening Portland cement	–	24	31	40
Sulfate resistance Portland cement	–	18	27	36
Low heat Portland cement	–	7	13	27
White Portland cement	–	18	27	36
Mixing sand Portland cement	–	12	20	27
4100 fines	10	25	32.5	40
Slag Portland cement	–	13	21	34
High-alumina cement				
80	25			
70	30			
50	50			
40	50			

As can be seen from Table 4.5, the high-alumina cement will gain the higher strength after 24 h. but on the other side there will have a high heat due to dehydration and in case of hot climate it can cause the cracks on concrete and this was happened on the past by using these types of cement.

It is very important to highlight that for low heat cement which is normally used in massive structure, you will have a low concrete strength after 3 days. Soremoving the shuttering shall be required more time than that in ordinary Portland cement and the strength will be lower also after 28 days so the concrete mix shall be different. So any mistake in concrete mix and use a different cement type will cause a lot of problems.

4.2.2 CEMENT STORAGE

In 1980 in Egypt there are a lot of building deficiency in strength and cracks. It is found by investigation that the main reason is the storage of the cement and the import cement. In most countries, there are some local cement factories and you can import from outside the country. The imported cement will be delivered through ship on the sea and also the silos were near the harbor for packing so the cement has a high humidity surrounding it. This makes a merge in the cement particles so you will loose the fines of the cement as we define from the previous laboratory test and you have a certificate for a good cement that matches with the standard but due to storing it will be out of the standard. On the other hand, there is some site mixing the concrete on the site by using cement bags so the storage of the cement bags is very important. I wish to highlight that the mixing design is based on cement match with the standards which provide you the required strength but actually due to bad storage so you will be under the specification requirement so you will not gain the required strength as per the concrete mix design.

Bad Materials Cause Failures

Figure 4.4 shows the relation between cement age and the storage condition and its effect on reduction of the cement mortar strength.

Figures 4.5 and 4.6 present the reduction in concrete strength depending on the storage condition with time. As a rule, it is prohibited to use cement over 30 days storage, as the concrete strength will depend on the days of storage.

As a conclusion the one of the main critical factor is the cement storage time period as per the above figures of storage condition one can see that, it can cause many dramatic problems on the structure as the strength will be reduce to around 50% at 180 days and from 35%–40% at 90 days.

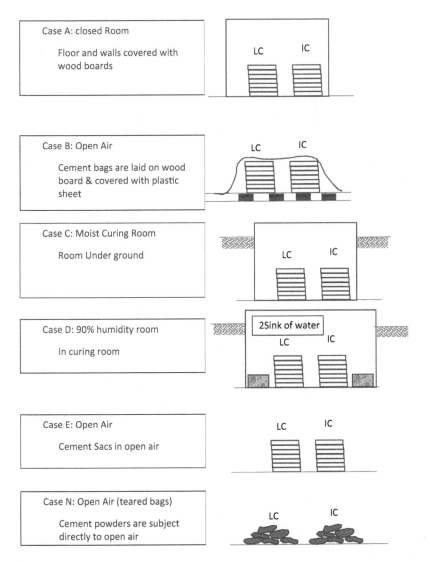

FIGURE 4.4 Cement storage.

58 Concrete, Steel, and Offshore Structures

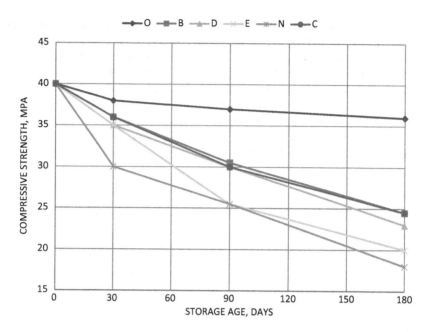

FIGURE 4.5 Effect of storage cases on 28 days compressive strength of mortar with local cement.

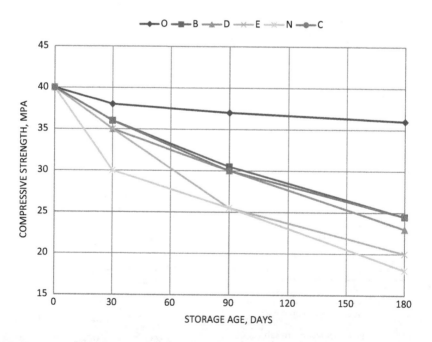

FIGURE 4.6 Effect of storage cases on 28 days compressive strength of mortar with Imported cement.

4.2.3 AGGREGATE TESTS

4.2.3.1 Sieve Analysis Test

The main key to obtain higher concrete strength is the aggregate interaction. This interaction increases the density of concrete, which increases the compressive strength. The interaction between aggregates depends on the grading of coarse aggregate and sand. So this grading should be according to definite specifications. This test relies on a standard sieve familiar with the specifications and dimensions, as shown in Table 4.6:

TABLE 4.6
Standard Sieve Sizes for Aggregates given by Various Standard (mm)

EC	BS410:1986	BS812:Section103.1:1985	EN 993-2	ASTM E11-87
	125			125
				100
	90			
75		75		75
63	63.0	63.0	63.0	
50		50		
	45			
37.5		37.5		37.5
	31.5			
		28		
26.5				
				25
	22			
				4
		20		
19				19
	16		16	
		14		
13.2				
				12.5
	11.2			
		10		
9.5				9.5
	8.0		8.0	
6.7				
		6.3		6.3
	5.6			
		5.0		
4.75				4.75
	4.0		4.0	
3.35		3.35		

(Continued)

TABLE 4.6 (*Continued*)
Standard Sieve Sizes for Aggregates given by Various Standard (mm)

EC	BS410:1986	BS812:Section103.1:1985	EN 993-2	ASTM E11-87
	2.8			
2.36		2.36		2.36
	2.0		2.0	
1.7		1.7		
	1.4			
1.18		1.18		1.18
	1.00		1.00	
0.85		0.85		
	0.710			
0.6		0.6		0.6
	0.5		0.5	
0.425				
	0.355			
0.3		0.30		0.30
	0.25		0.25	
0.212		0.212		
	0.18			
0.15		0.15		0.15
	0.125		0.125	
	0.090			
0.075		0.075		0.075
	0.063		0.063	
	0.045			
	0.032			

These standard sieves are a metal cylinder frame with a square opening. The sieve is classified as per its opening length in millimeter and the sieve shape, as shown in Figure 4.7.

The test procedure starts by defining weight of the aggregate, and then drying it at a temperature of 105°C ± 5°C for 24 h until the weight proves to the nearest 0.1%.

Sieves are arranged in the order according to size, and then put the sample on the largest size of the sieves. Start shaking the sieve manually or mechanically for sufficient period not less than 5 min so as to ensure that no more than 0.1% of the total weight of the sample passes through during 1 min of manual shaking.

Take into account only pay granules aggregate to pass through Sieve by applying hand pressure, but you can do it only for sieve size 20 mm and more.

Then, measure the weight of the remaining aggregate on the sieves separately and then calculating that which passed through the sieves, as shown in the Table 4.7.

Bad Materials Cause Failures

FIGURE 4.7 Shapes of sieves.

TABLE 4.7
Calculate the Percentage of Remaining and Passing from the Sieve Analysis

Sieve Size (mm)	Remaining Weight on the Sieve	Total Remaining Weight on the Sieve	Percentage of Remaining Aggregate	Percentage of Passing Aggregate
37.5	W_1	W_1	$R_1 = W_1/W$	$100 - R_1$
20	W_2	$W_1 + W_2$	$R_2 = (W_1 + W_2)/W$	$100 - R_2$
10	W_3	$W_1 + W_2 + W_3$	$R_3 = (W_1 + W_2 + W_3)/W$	$100 - R_3$
5	W_4	$W_1 + W_2 + W_3 + W_4$	$R_4 = (W_1 + W_2 + W_3 + W_4)/W$	$100 - R_4$

The nominal maximum aggregate size is defined as the smallest sieve that passes at least 95% of the coarse aggregate or whole aggregate.

During this test, do not put over weight on the sieve as the maximum weight of the remaining aggregate on the sieve should not exceed the weights in Table 4.8, which is based on ECP code and BS 812: section 103.1;1985.

TABLE 4.8
The Maximum Weight for Remaining Aggregate in Different Sieve Size

Sieve Opening Size (mm)	Max. Weight (kg) Sieve Diameter 450 (mm)	Sieve Diameter 300 (m)	Sieve Opening Size (mm)	Max. Weight (kg) Sieve Diameter 300 (m)	Sieve Diameter 200 (m)
50	14	5	5.00	750	350
37.5	10	4	3.35	550	250
28	8	3	2.36	450	200
20	6	2.5	1.80	375	150
14	4	2	1.18	300	125
10	3	1.5	0.85	260	115
6.3	2	1	0.6	225	100
5	1.5	0.75	0.425	180	80
3.35	1	0.55	0.300	150	65
			0.212	130	60
			0.15	110	50
			0.075	75	30

The above table based on ISO ISO6274-1982. The acceptable and refusal limits for the coarse aggregate and sand, and the whole aggregate are based on ECP codes, BS882:1992 Nd ASTMC33-93, as shown in Tables 4.9 and 4.10:

The following table is the grading requirement of coarse aggregate according to the BS882:1992 which is similar with the ECP but to some modification (Tables 4.11, 4.12, and 4.13).

The acceptable grade requirement in British standard BS882;1973 is shown in Table 4.14.

The actual grading requirements depend to some extent on the shape and surface characteristics of the particles. For instance, sharp, angular particles with rough surfaces should have a slightly finer grading in order to reduce the interlocking and

TABLE 4.9
Acceptable and Refusal Limits for Fine Aggregate

Sieve Opening Size (mm)	Percentage of Passing the Sieve				ASTM C33-93
	ECP and BS882;1992				
	General Grading	Coarse	Medium	Fine	
10.0	100	–	–	–	100
5.0	89–100	–	–	–	95–100
2.36	60–100	60–100	65–100	80–100	80–100
1.18	30–100	30–90	45–100	75–100	50–85
0.6	15–100	15–45	25–80	55–100	25–60
0.3	5–70	5–40	5–48	5–70	10–30
0.15	0–15	–	–	–	2–10

Bad Materials Cause Failures

TABLE 4.10
Acceptable and Refusal Limits for Coarse Aggregate

| | Percentage by Weight Passing Sieves |||||||
| | Nominal Size of Graded Aggregate (mm) ||| Nominal Size of Single-Sized Aggregate Size (mm) ||||
Sieve Opening Size (mm)	5–40	5–20	5–10	40	20	14	10
50.0	100	–	–	100	–	–	–
37.5	90–100	100	–	85–100	100	–	–
20.0	35–70	90–100	100	0–25	85–100	100	–
14.0	–	–	90–100	–	–	85–100	–
10.0	10–40	30–60	50–85	0–5	0–25	0–50	100
5.0	0–5	0–10	0–10	–	0–5	0–10	50–100
2.36	–	–	–	–	–	–	0–30

TABLE 4.11
Acceptable and Refusal Limits for Coarse Aggregate in BS882:1992

| | Percentage by Weight Passing Sieves |||||||
| | Nominal Size of Graded Aggregate (mm) ||| Nominal Size of Single-Sized Aggregate Size (mm) ||||
Sieve Opening Size (mm)	5–40	5–20	5–14	40	20	14	10
50.0	100	–	–	100	–	–	–
37.5	90–100	100	–	85–100	100	–	–
20.0	35–70	90–100	100	0–25	85–100	100	–
14.0	25–55	40–80	90–100	–	0–70	85–100	100
10.0	10–40	30–60	50–85	0–5	0–25	0–50	85–100
5.0	0–5	0–10	0–10	–	0–5	0–10	0–25
2.36	–	–	–	–	–	–	0–5

to compensate for the high friction between the particles. The actual grading of crushed aggregate is affected primarily by the type of crushing plant employed. A roll granulator usually produces fewer fines than other types of crushers, but the grading depends also on the amount of material fed into the crusher.

4.2.3.2 Abrasion Resistance of Coarse Aggregates in Los Angeles Test

It is usual to see some deterioration on the surface on the slab on grade and specific if there are cars travelling on it regularly like garage or warehouse. These degradation is responsible for the abrasion on the slab surface and this happened as no abrasion test was done to the aggregate for these types of structure. As we discussed before, the civil engineers on the site focus on the compression test and slump test which cannot be predicted if the aggregate has a reasonable abrasion resistance based on the standard or not.

TABLE 4.12
Grading Requirements for Coarse Aggregate According to ASTM C33-93

	Percentage by Weight Passing Sieves				
	Nominal Size of Graded Aggregate (mm)			Nominal Size of Single-Sized Aggregate Size (mm)	
Sieve Opening Size (mm)	37.5–4.75 mm	19.0–4.75 mm	12.5–4.75 mm	63	37.5
75	–	–	–	100	–
63	–	–	–	90–100	–
50.0	100	–	–	35–70	100
38.1	95–100	–	–	0–15	90–100
25	–	100	–	–	20–55
19	35–70	90–100	100	0–5	0–15
12.5	–	–	90–100	–	–
9.5	10–30	20–55	40–70	–	0–5
4.75	0–5	0–10	0–15	–	–
2.36	–	0–5	0–5	–	–

TABLE 4.13
Acceptable and Refusal Limits or Whole Aggregate According to BS882:1992 and ECP2002

	Percentage of Passing the Sieve		
Sieve Opening Size (mm)	Nominal Maximum Aggregate Size 40 mm	Nominal Maximum Aggregate size 20 mm	Nominal Maximum Aggregate size 10 mm
50.0	100	–	–
37.5	95–100	100	–
20.0	45–80	95–100	–
14.0	–	–	100
10.0	–	–	95–100
5.0	25–50	35–55	30–65
2.36	–	–	20–50
1.18	–	–	15–40
0.60	8–30	35–10	10–30
0.30	–	–	5–15
0.15	[a]0–8	[a]0–8	[a]0–8

[a] Increase to 10% for crushed rock fine aggregate.

From this test, one can define the abrasion factor, which is the percentage of weight loss due to abrasion in Los Angeles test from its original weight.

Los Angeles device is a cylinder that is rotated manually. Inside it are balls made from cast iron or steel with diameter of about 48 mm and weight per ball ranged between 3.82 and 4.36 N.

TABLE 4.14
Grading Requirement for Whole Aggregate Based on BS882:73

| Sieve Size (mm) | Percentage by Weight Passing Sieves ||
	Nominal Size 40 mm	Nominal Size 20 mm
75	100	–
37.5	95–100	100
20	45–80	95–100
5	25–50	35–55
600 μm	8–30	10–35
150 μm	0–6	0–6

This test begins by bringing a sample of big aggregate, weighing from 5 to 10 kg, and washing the sample by water and then drying it in an oven at a temperature of 105°C–110°C until it reaches proved weight.

Separate the samples into different sizes through the sieves shown in Table 4.15. Then collect the test sample from the aggregate by mixing the weights.

The sample will be weighed after remixing it. The weight is (W_1) and the grading type is as per the table from A to G. By knowing the grading type, define the number of bars that will be put in the device, as shown in Table 4.16.

Put the sample and the balls inside the Los Angeles machine and rotate the machine with speed 10–31 rpm so that the total number of rotations is 500 for sample gradients, A, B, C, D, F and 1000 cycle for the rest of grading.

Lift aggregate from the machine and put in sieve size 16 mm and then pass it through sieve size 1.7 mm.

Then wash the aggregates that remain on the two sieves, and then dry in oven under temperature 105°C–110°C and then weight it to be (W_2).

$$\text{The percentage of abrasion} = \frac{W_1 - W_2}{W_1} \times 100$$

TABLE 4.15
Collect Sample Test after Sieve Analysis

Passing From	Remaining On	A	B	C	D	E	F	G
75.00	63.00					1,500		
63.00	50.00					1,500		
50.00	37.5					1,500	5,000	
37.5	25.00	1,250					5,000	5,000
25.00	19.00	1,250						5,000
19.00	12.5	1,150	1,500					
12.5	9.5	1,150	1,500					
9.5	6.3			1,500				
6.3	4.75			1,500				
4.75	2.38				5,000			

TABLE 4.16
Define the Number of Abrasion Ball

A	13
B	11
C	8
D	6
E	11
F	11
G	11

The acceptable aggregate that the percentage of abrasion by Los Angeles test is not more than 20% in aggregate and 30% for crushed stone.

4.2.3.3 Determination of Clay and Other Fine Materials in Aggregates

The clay and fine materials that pass through micron-sized sieves are known as soft material. Therefore, we are conducting this test to make sure that aggregate is conformity with the standard specifications. This will be done by taking a fine aggregate sample of at least 250 g. In the case of coarse aggregate or whole aggregate, the sample weight will be as in Table 4.17.

Test begins by drying the sample in an oven at a temperature of 110°C ± 5°C until it reaches proved weight (W). Then immerse the sample in water and move it strongly. Remove the clay and fine materials from the aggregate sample by putting the wash water on the two sieves No. 75 micron and No. 141 micron. The No. 141 micron will be on the top. Then repeat the washing several times until the washing water is pure.

Return the remaining materials on sieve 141 and 75 micron to the washed sample. then dry the remainder at the same oven temperature until the weight W_1. Then calculate the percentage of clay and fine materials by weight as follows:

$$\text{Percentage of fine materials and clay} = \frac{W - W_1}{W} \times 100$$

Based on ASTM C142-78, and BS882:1992 specifications, the maximum acceptable limits for clay and fine materials in aggregate are shown in Table 4.18.

TABLE 4.17
Sample Weight to Test Percentage of Clay and Fine Aggregate

Max. Nominal Aggregate Size (mm)	Least Sample Weight (kg)
9.5–4.75	5
19–9.5	15
37.5–19	25
<37.5	50

TABLE 4.18
Maximum Allowable Limits or Fine Aggregate in Concrete

Type of Aggregate	Percentage of Clay and Fine Materials by Weight (%)
Sand	3
Fine aggregate by crushing stone	5
Coarse aggregate	1
Coarse aggregate from crushed stone	3

4.2.3.3.1 On Site Test

This simple test can be conducted on the with 50 cm^3 of the clean water filling a glass tube and then adding the quantity of sand until the total volume is 100 cm^3. Then add clean water until the total volume is 150 cm^3 as shown in Figure 4.8.

Strongly shake the glass tube until the fine materials and clay move up. Set the tube on a horizontal table for 3 h. Then calculate the percentage of fine materials and clay by its volume that hangs on the top layer of the water with respect to the volume of the aggregate that has settled to the bottom of the tube.

So if you find a higher percentage of suspended clay and fine materials, the aggregate is not a match with the specification. You should go to the previous laboratory test, which is the official test to accept or refuse the aggregate.

4.2.3.4 Aggregate Specific Gravity Test

The specific gravity of the aggregate is apparent density relative to the rule of solid constituent and there are no spaces inside it to access the water.

The specific gravity is obtained by dividing the weight of dry aggregate by water volume (Table 4.19).

4.2.3.5 Fine Aggregate Test

Test sample is determined so as not to exceed 100 g. Then dry it in the ventilated oven temperature from 100°C to 110°C. Cool the sample in the dryer, weigh it, return it to the drying process, and weigh it until it proves weight and measurement (W).

FIGURE 4.8 Tubes for aggregate density test.

TABLE 4.19
Aggregates Specific Gravities

Type of Aggregate	Range of Specific Gravity
Sand	2.5–2.75
Coarse aggregate	2.5–2.75
Granite	2.6–2.8
Basalt	2.6–2.8
Limestone	2.6–2.8

Add the water at a temperature of 15°C–25°C to a graduated tube and then add the fine aggregate and leave it submerged in water for an hour. Remove any bubbles by knocking on the tube.

Read the level of the water in the tube after 1 h adding the aggregates. Then define the aggregates volume (V) and then calculate the specific gravity by the following equation:

$$\text{Specific gravity of aggregate} = \frac{W}{V}$$

4.2.3.6 Define Specific Gravity for Coarse Aggregate

Immerse a sample of about 2 kg in a water with a temperature 15°C–12°C for 24 h. Then take the aggregate and dry it manually with a piece of wool.

Put a certain volume of water in the big bowl to its midpoint with water. This is volume (V_1).

Then add the more water to fill the bowl completely. This identifies volume (V_2) and then take the aggregate, dry it, and measure its weight (W). Calculate the coarse aggregate specific gravity by the following equation:

$$\text{Coarse aggregate specific gravity } (S_g) = \frac{W}{V_2 - V_1}$$

4.2.3.7 Bulk Density or Volumetric Weight Test for Aggregate

This test determines volumetric weight of the aggregate. By knowing the volumetric weight we can transform a given volume of aggregate equivalent to the weight or vice versa. By knowing the volumetric weight and specific gravity, we can calculate the percentage of voids between aggregate grains.

The definition of volumetric weight is the ratio of the weight of the aggregate to its volume that occupies. The percentage of voids is the ratio between voids between the aggregate and the total volume of the aggregate that occupy.

Define the bowl or the test based on Table 4.20.

By knowing the volume of the bowl (V), measure the bowl weight when it is empty (W_1). Fill the container with the aggregate and perform standard rod compaction by 25 times and do it twice until the bowl is completely full. Then measure the weight of the bowl with the compacted aggregate (W_2).

The volumetric weight can be calculated by the following equation:

$$\text{Volumetric weight } (V_w) = \frac{W_2 - W_1}{V}$$

Bad Materials Cause Failures

TABLE 4.20
Sizes of Containers for Define Aggregate Bulk Density

Max. Nominal Aggregate Size (mm)	Bowl Volume (L)	Bowl Dimensions (mm) Internal Diameter	Internal Height	Thickness
>40	30	360	293.6	5.4
40-5	15	360	282.4	4.1
<5	3	155	158.9	3.0

$$\text{Void percentage} = \frac{V_w - S_g}{V_w} \times 100$$

4.2.3.8 The Percentage of Aggregate Absorption

This test is used to identify the absorption of water by the aggregate with max nominal size higher than 5 mm and the following table is the limits of absorbing aggregate to the water that should be follow in selecting the aggregate before the concrete mix as it will affect the percentage of water in the concrete mix (Table 4.21).

Beginning of the test is determined weight about 100 times the maximum nominal aggregate size in mm. washing the sample before test on sieve 5 mm to remove the suspended materials.

Put the sample in a wire mesh of 1–3 mm and then immersed in the container full of water at a constant temperature of 15°C–25°C so that the total immersion in less distance between the highest point in the basket and surface water on 50 mm.

After immersion remove the entrained air and leave the basket and sample immersing for 24 h.

Then took place the basket and the sample then remove the suspended water on them. Dry the surfaces of the sample gently and distribute it in piece of cloth and left it in the air and away from the sun or any source of heat and weight it (W_1).

Then put the sample in oven for temperature 105°C ± 5°C for 24 h and then let it cool without expose to any humidity and weight it (W_2).

$$\text{The absorption percentage} = \frac{W_1 - W_2}{W_2} \times 100$$

TABLE 4.21
Maximum Allowable Limit of Absorption Water by Aggregate

Type of Aggregate	Percentage of Absorption (%)
Quartzite and crushed limestone	1–0.5
Granite	0–1
Stone	Not more than 2.5

TABLE 4.22
Maximum Allowable Limits to Salt and Suspended Materials in Water

Type of Salt and Suspended Materials	Maximum Limit to Salt Content (g/L)
Total dissolved solid (T.D.S)	2.0
Chloride salt	0.5
Sulfate salt	0.3
Carbonate and bicarbonate salt	1.0
Sodium sulfate	0.1
Organic materials	0.2
Clay and suspended materials	3.0

4.2.4 Mixing Water Test

This test is very mandatory as it defines the chloride content, if you will not doing this test so you can a have a lot of durability problem of your structure and really a lot of problems in cracks due to corrosion. It is important to highlight that over 40 years it was not prohibited to use a salt water as it increases the concrete strength but with time we have a lot of problems worldwide due to this approach.

The water is very essential factor for mixing the concrete and the curing process so the quality of water is essential to concrete durability as it must follow the project speciation's.

Table 4.22 illustrates the acceptable water specification for mixing and curing process.

The determination of soluble salts in water can be obtained by taking a sample of approximately 25 mL of water nominated before and placed in dish platinum. Evaporate the sample and then transfer to the drying oven at 105° until fix the measured weight. By knowing size of the sample, we obtain the total content of dissolved salts in water (TDS) (g/L).

Chloride content in water is defined by performing chemical analysis test to compare the content of chloride of water with the permissible content in the specifications as in the above table.

4.3 ADMIXTURES

Now admixtures are widely used in concrete industry as it increases the concrete compressive strength or controls the rate of hardening the fresh concrete or increases the concrete workability.

These admixtures are powder or liquid materials which are produced from carbohydrate, melamine, condensate, naphthalene, and organic and non-organic materials.

The type of admixture should be identified correctly and also the dose that is required to achieve the target for using this admixture.

There are some tests that should be applied to the admixture before use and compare it with the acceptable and refusal limits based on the project specification.

The types of admixtures are as follows:

Bad Materials Cause Failures

1. Normal setting water reducer

   This admixture increases the workability during concrete mixing without any change in water–cement ratio or maintain the workability and decrease the water–cement ratio, so it will increase the concrete strength and this type will be matched with ASTM C494, type A-Normal Setting ASTM C494 type A.

2. Retarder

   These admixtures reduce the rate of reaction between cement and water so it will retard the concrete setting and hardening. This type of admixtures follows ASTM C 494, Types B, and D.

   This additive is usually used in ready-mix concrete to achieve the required time to reach the site before setting the concrete so it need a retarder to the setting time.

3. Accelerators

   These admixtures increase the rate of chemical reaction between cement and water that increase the rate of setting and hardening concrete and it follows ASTM C 494, Types C, and E.

   Accelerators are not normally used in concrete unless early form removal is critical and important to project execution plan.

   Accelerators are added to concrete either to increase the rate of early strength development or to shorten the time of setting. The advantages are as follow:
   a. earlier removal of forms
   b. reduction of curing time
   c. early usage of the structure
   d. partial compensation for the effects of low temperature on rate of strength development.
   e. early finishing of surfaces
   f. reduction of pressure on forms

4. Admixture for water reducer and retarder

   These admixtures combine between reduce water content by increase the workability and retarding the setting time in the same time.

5. Admixture for water reducer and accelerator

   These admixtures combine between reduce water content by increase the workability and accelerate the setting time in the same time hence we obtain two functions by same additive.

6. High-range water reducer

   These are modern type of water-reducing admixture, much more effective than the admixtures that is discussed in the preceding items.

   Based on Nevil at a given water/cement ratio, this dispersing action increases the workability of concrete, typically by raining the slump from 75 mm to 200 mm, the mix remaining cohesive. Note that the improvement in workability is smaller at high temperatures.

7. High-range water reducer and retarder

   These additives increase the durability by reducing water cement ratio so it will increase the concrete compressive strength and in the same time it will increase the setting time.

4.3.1 Samples for Test

The admixtures are found as liquid or powder so the sample should be taken in the proper matter to represent the deliverables quantity on site.

a. Powder admixtures

In this state, the sample will be taken to present not more than or equal 1 ton and the samples is taken from six packs or from 1% of the total number of packs or from all packs if the number of packs not more than six. The samples packs should represent all the packs.

b. Liquid admixtures

The samples will be taken from six drums or 1% from the number of all drums whatever larger or take from all the drums if the number of drums less than six. These samples represent an order which is not more than 5000 L from the liquid admixtures. Shake the drums to distribute the suspended materials and ignore the remaining sediments after shaking.

4.3.2 Chemical Tests to Verify Requirements

The chemical test should be performed to the admixtures to measure the following criteria.

4.3.2.1 Chemical Tests

The chemical test is performed to measure some parameters and compare results with the product data sheet to determine if a sample matches with manufacturer specifications.

A. For admixtures in the form of powder, the humidity will be removed from it by weight of about 3 g of the admixtures and remove the humidity and then determine of the percentage of the content of solid material.

B. For liquid admixtures,
- It is required to put from 25 to 30 g of sand passing through sieve No. 30 into a glass bottle with rough surface opening with internal diameter 60 mm, height 30 mm and a cover of the provisions of closure.
- Put the bottle and the cover in the drying oven at 105°C–110°C and leave for 17 h ± 15 min.
- Cover the bottle and place it in the dryer until it reaches room temperature and then weigh it to the nearest 0.001 gm and note the weight (W).
- Place about 4 mL from the sample inside the bottle over the sand and weigh it to the nearest 0.001 and note the weight (W_1).
- Put the bottle in the dryer oven at the same temperature for 17 h + 15 min.
- The bottle cover and place in the dryer at room temperature and weigh it to the nearest 0.001 g and note the weight (W_2).

The percentage of the solid materials will be calculated from the following equation:

$$\text{Percentage of solid content} = \frac{W - W_2}{W - W_1} \times 100$$

Bad Materials Cause Failures

4.3.2.2 Ash Content

The purpose of this test was to determine the content of non-organic materials through the analysis of ash content. This test is summarized in the following steps:

1. Heat the container with its cover at a temperature of 600°C for 15–30 min and then transfer to dryer.
2. Now cool for 30 min and then weigh with the cover. The weight is (W_1).
3. Add about 1 g of the required admixture to test and then re-cover and measure this weight (W_2).
4. To reduce mechanical heat, spray the sample with water and keep in 90°C drying oven.
5. Heat the sample to 300°C for an hour and then increase heat to 600°C for 2–3 h. Now leave the sample at a temperature of 600°C for 16 ± 2 h.
6. Transfer to drying oven and allow to cool with the cover. Weigh the content to the nearest 0.001 g after 30 min of cooling. Its weight is (W_3).

The ash content value will be calculated from the following equation:

$$\% \text{ Ash content} = \frac{W_1 - W_3}{W_1 - W_2} \times 100$$

where w_1 is the weight of container + cover, W_2 is the weight of container + cover + sample weight before burning, and W_3 is the weight of container + cover + sample weight after burning.

4.3.2.3 Relative Density

In this test, put a sample at a temperature of $20°C \pm 5°C$ and then transfer the admixture sample to tube with a capacity of 500 mL of the hydrometer immersed in the liquid inside tube. Then leave the hydrometer to reach the balance and then read the value hydrometer, then read the grading at the base of the surface of contact with the liquid. Record the density to the nearest 0.002.

4.3.2.4 Define the Hydrogen Number

It will be used a special apparatus to define the number of hydrogen, to define the pH value of a sample. in the case of admixtures in the shape of powders to be prepared in the form of liquid for the appointment of pH test and compared with the specifications of the product.

4.3.2.5 Define Chloride Ion

In this test, prepare two equal standard solutions of sodium chloride and then add each sample solution to the admixture sample. Estimate the proportion of chlorides after each addition and calibrate it with a standard solution of silver nitrate using a pole of silver to determine the equal point of differential voltage.

Thus, we can assess the chloride ion content in the samples containing a very small percentage of chloride and can be at the same time estimate the standard silver nitrate and control the quantity of required sodium chloride.

Note that any method can be used to determine the chloride content and have the same accuracy (Table 4.23).

TABLE 4.23
Admixture Characteristics

Character	Requirement
Solid content	Not more than 5% of the weight on the value state by the manufacturer for solid and liquid admixtures
Ash content	Not more than 1% of weight for the value stated by the manufacturer
pH number	Compare by the number defined by the manufacture
Chloride ion content	Not more that 5% for the value stated by the manufacturer or 0.2% from the weight whatever higher

4.3.3 Performance Tests

The purpose of these tests is to identify the performance of admixtures and the degree of influence over the conduct tests on fresh and hardened concrete and compare the same concrete mixtures without admixtures.

To determine the performance of the admixtures, the test will be done for two mixtures which have the same specification but one of them contains admixtures. The mixture without admixtures is called the control specimen.

The admixture samples to be tested are obtained in a liquid state or in the form of powders, as mentioned earlier.

4.3.3.1 Control Mixing

The control mixing consists of an ordinary Portland cement. The coarse aggregate must be identical to the standard specifications, but fully dry by using the oven, and be clean and free of organic substances and impurities. The coarse aggregates have two sizes 20–10 mm and second size 10–5 mm. The flakiness index must not exceed 35%.

The sand follows the same specifications of the coarse aggregate and also must be dried using the oven, and acid dissolved by no more than 5%.

1. Ratio of specimen control without admixture
   - Cement $300 \pm 5 \, kg/m^3$
   - Cement to whole aggregate ratio 1:6 by weight.
   - Percentage of coarse aggregate by weight: 45% for sizes 10–20 mm, 20% for sizes 10–5 mm and 35% for sand.
   - Define the water quantity to provide slump 60 ± 1 mm or compaction ratio between 88% and 94%
   - Air entrained is not more than 3%.
2. Ratio of specimen control with admixture

    Use the same ratios in the case of mixing control but add the admixtures with the same ratio as specified by the manufacturer, taking into account the content of the water that makes concrete with the same mix of operational control. The content of the air entrained in the control specimen should not be more than 2% and the total content of the air is not more than 3%.

    Table 4.24 presents a comparison between the performance of the concrete by adding admixtures and the control mix without adding the admixtures.

Bad Materials Cause Failures

TABLE 4.24
Performance Requirement or Concrete with Admixtures

				Admixture Type				
		WR[a]	Retarder	Accelerator	WR & Retarder	WR & Accelerator	HWR[a]	HWR & Retarder
Maximum water content as percentage from mix control		95	—	—	95	95	88	88
Air content		Not increaser more than 25 in concrete with admixture than the control specimen without admixtures and the total air content must be not more than 3%						
Hardening time at penetration resistance	0.5 N/mm²	During 1 h from SC[b]	—	—	1 h more than SC as lower limit	More than 1 h	—	—
	3.5 N/mm²	During 1 h from SC	—	—	—	1 h lower than SC as lower limit	—	—
Setting time at penetration resistance	3.5 N/mm²	During 1 h from SC	From 1 to 3 h more than SC	From 1 to 3 h lower than Sc and not less than 45 min.	From 1 to 3 h more than SC	From 1 to 3 h lower than Sc and not less than 45 min.	During 1 h from SC[b]	From 1 to 3 h more than SC
	27.6 N/mm²	During 1 h from SC	Until 3 h more than SC	Until 1 h less than SC and not less than 45 min.	Until 3 h more than SC	Until 1 h less than SC and not less than 45 min.	During 1 h from SC[b]	Until 3 h from the SC

[a] WR is the water reducer and HWR is the higher water reducer.
[b] SC is the specimen control mix.
[c] Setting time according to ASTM C 494-1996.

TABLE 4.25
Relation between Type of Admixtures and Minimum Limit to Concrete Strength as Percentage from Specimen Control Mix

Age, Days	WR	Retarder	Accelerator	WR& Retarder	WR& Accelerator	HWR	HWR & Retarder
1	–	–	125	–	125	140	125
3	110	90	125	110	125	125	125
7	110	90	100	110	110	115	115
28	110	90	100	110	110	110	110
180	100	90	90	100	100	100	100

This table provides the guide to accept or refuse the admixtures based on its function. For example, a retarder giving a retarding time only 30 min should be refused as the retarding time must be from 1 to 3 h.

The relation between the type of admixtures and the minimum acceptable limits to the concrete compressive strength as a percentage of the concrete compressive strength in control mix without admixtures is shown in Table 4.25.

Table 4.25 is very important as it present the root cause if there are any failure in case of using concrete retarder as it will decrease the concrete strength by 10% after 28 days so adding the water reducer is very important with the retarder. This table presents also that after 6 months the concrete strength will reduce 10% in case of using accelerator so after many years approximately there will be no gain on strength rather than the concrete without using the accelerator.

4.4 STEEL REINFORCEMENT TEST

The steel reinforcement is an important element of reinforced concrete structures where the tensile loads are the responsibility of reinforcing steel bars in reinforced concrete. Therefore, we must make sure that rebar meet specifications for the project strictly. Therefore, we will present here some important tests for quality control rebar supplier to the project.

4.4.1 Weights and Measurement Test

This test requires a measuring tape as well as the relevant Vernier.

The samples will withdraw two samples of the same diameter of each consignment weighing less than 50 tons. Withdraw three samples of each country if the consignment weighing is more than 50 tons.

Sure measuring bar diameter. In each sample, measure two perpendicular diameters in the same cross-section by using a special measurement unit (Table 4.26).

To ascertain the weight of longitudinal meter a sample with length not less than 500 mm with accuracy ± 0.5% is required.

The actual cross-sectional area, A, is calculated as follows:

TABLE 4.26
Weight of the Deformed Bar per Unit Length

Nominal Diameter (mm)	Nominal Cross Sectional Area (mm²)	Weight (kg/m)	Tolerance Allowance (%)
6	28.3	0.222	±8
8	50.3	0.395	
10	78.5	0.618	±5
12	113	0.888	
13	133	1.04	
14	154	1.21	
16	201	1.58	
18	254	2.00	
19	283	2.22	
20	314	2.47	
22	380	2.98	
25	461	3.85	±4
28	616	4.83	
32	804	6.31	
36	1020	7.99	
40	1257	9.86	
50	1964	15.41	

$$\text{Actual cross sectional area} = \frac{W}{0.00785L}$$

where
  W is the weight to nearest kg and
  L is the length to nearest mm.

4.4.2 Tension Test

This test is useful to define the mechanical properties of steel bars to expose the test sample with the standard dimensions to tensile stress until fracture. Figure 4.9 presents the shape of the tension machine and Figure 4.10 presents the elongation measurement. From this test, we can define the yield stress or proof stress and also the elongation percentage:

$$\text{Elongation percentage} = \frac{L_2 - L_1}{L_1} \times 100$$

where
  L_1 is the original measured length and
  L_2 is the final measured length.

The length of the short sample is $L_1 = 5D$ and the length of the long sample is $L_1 = 10D$, where D is the diameter of the steel bar and the deviation for the measurement of sample dimensions is not more than ± 0.5%.

FIGURE 4.9 Steel bar in a tension machine.

FIGURE 4.10 Elongation measurement.

During sample preparation, we can do some modification but it is prohibited to change the shape of the sample by increasing its temperature as exposing the sample to heat:

$$\text{Yield stress} = \frac{\text{yield force}}{\text{actual cross sectional area}}$$

$$\text{Tension strength} = \frac{\text{maximum force}}{\text{actual cross sectional area}}$$

The Egyptian standard provides specifications that are required to achieve at least 95% of the quantity tested values set in Table 4.27. Moreover, the result of any single test not less than 95% is also presented in the table. It can be agreed between the supplier and customer to ensure that the values shown in the table above are the minimum acceptance limits to the bars.

The ratio between the tension strength and yield strength for any sample is not less than 1.1 and 1.05 for smooth and ripped bar, respectively. Steel that cannot define the yield point clearly defines the proof stress 0.02% instead of yield stress.

TABLE 4.27
Acceptable and Refusal or Mechanical Properties

Steel Grade	Minimum Yield Strength (N/mm^2)	Minimum Tensile Strength (N/mm^2)	Elongation (%)
240	240	350	20
280	280	450	18
360	360	520	12
400	400	600	10

If there are any reduction on the yield strength or execute by steel with wrong grade with different yield strength rather than the design one, it will cause a structure failure but if there are a reduction on the steel bar diameter it will cause a problem along the structure life time. The elongation is very important as the capability of the structure to withstand the lateral load like the earthquake or wind load it depends on the steel amount and its ductility which present here in the test by the elongation percentage. Note that the factors of safety and design calculation percentage consider this elongation based on the steel grade so if the delivered steel on site have less amount of ductility it main a serious failure or major cracks after earthquake effect.

BIBLIOGRAPHY

ACI 228.1R89, In-place methods or determination of strength o concrete, ACI manual of concrete practice, part 2: construction practices and inspection pavements, 25 pp. (Detroit, Michigan, 1994).
ASTM C188-84, Density of hydraulic cement.
ASTM C114-88, Chemical analysis of hydraulic cement.
ASTM C183-88, Sampling and amount of testing of hydraulic cement.
ASTM C349-82, Compressive strength of Hydraulic cement Mortars.
ASTM C670-84, Testing of building materials.
ASTM C142-78, Test method for clay lumps and friable particles in aggregate.
ASTM D1888-78, Standard test method for particulate and dissolved matter, solids or residue in water.
ASTM D512-85, Standard test method for chloride ion in water.
ASTM D516-82, Standard test method for sulfate ion in water.
Ajdukiewicz, A.B., and Kliszczewicz, A.T, Utilization of recycled aggregates in HS/HPC, *5th International Symposium on Utilization of High strength/High Performance Concrete*, Sandefjord, Norway, June 1999, Vol. 2, pp. 973–980, 1999.
Ajdukiewicz, A.B., and Kliszczewicz, A.T., Properties and usability of HPC with recycled aggregates. *Proceeding, PCI/FHWA/FIP International Symposium on High Performance Concrete*, Sep., 2000, Orlando, Precast/Prestressed Concrete Institute, Chicago 2000, pp. 89–98, 2000.
Ajdukiewicz, A.B., and Kliszczewicz, A.T., Behavior of RC beams from recycled aggregate concrete, *ACI 5th International Conference*, Cancun, Mexico, 2002.
BS 882, 1992.
BS 812 Part 103-19, Sampling and testing of mineral aggregate sands and fillers.
BS 410-1:1986, Spec. for test sieves of metal wire cloth, 2000.

Bs EN933-1 1997, Tests for geoetrical properties of aggregates, determination of particle size distribution. Sieving method.

Di Niro, G., Dolara, E., and Cairns, R., The use of recycled aggregate concrete for structural purposes in prefabrication. *Proceedings, 13th FIP Congress "Challenges for CONCRETE IN THE Next Millennium"*, Amsterdam, June 1998; Balkema, Rotterdam-Brookfield, 1998, V.2, pp. 547–550, 1998.

Egyptian Standard Specification, 1947-1991, method of taking cement sample.

Egyptian Standard Specification, 2421-1993 natural and mechanical properties for cement. Part 2: define cement finening by sieve No.170.

Egyptian Standard Specification, 2421-1993 natural and mechanical properties for cement. Part 2: define cement finening by using Blaine apparatus.

Egyptian Standard Specification, 2421-1993 natural and mechanical properties for cement. Part 1: define cement setting time.

Egyptian Standard Specification, 1450-1979 Portland cement with fines 4100.

Egyptian Standard Specification, 1109-1971 concrete aggregate from natural resources.

Egyptian Standard Specification, 262-1999 steel reinforcement bars.

Egyptian Standard Specification, 76-1989 tension tests for metal.

Egyptian code for design and execute concrete structures: part3 laboratory test for concrete materials, ECP203, 2003.

El-Arian, A., El-hakim, F., and Abd El-Aziz, M.I., "Effect of storage condition of ordinary Portland cement on its properties", M.Sc., Structural dept., Cairo University, 1985.

Elreedy, M.A., "New project management approach for offshore facilities rehabilitation projects", SPE-160794, *Adipec Abu Dhabi Conference proceeding*, UAE, 2012.

ISO 6274-1982, Sieve analysis of aggregate.

Kasai, Y., (ed.), Demolition and reuse of concrete and masonry reuse of demolition waste, *Proceedings, 2nd International Symposium RILEM*, Building Research Institute and Nihon University, Tokyo, Nov., 1988 Chapman and Hall, London-New York, pp. 774, 1988.

Mukai T., and Kikuchi, M., Properties of reinforced concrete beams containing recycled aggregate. *Proceedings, 2nd International Symposium RILEM "Demolition and Reuse of Concrete and Masonry"*, Tokyo, Nov., Chapman and Hall, London-New York, Vol. 2, pp. 670–679, 1988.

Murphy, W.E., discussion on paper by Malhotra, V.M., 1977. Contract strength requirements-core versus in situ evaluation. *ACI Journal, Proceedings*, Vol. 74, No. 10, pp. 523–5.

Neville, A. M., Properties of concrete, PITMAN, 1983.

Plowman, J.M., Smith, W.F., and Sheriff, T., 1974. Cores, cubes, and the specified strength of concrete, *The Structural Engineer*, Vol. 52, No. 11, pp. 421–6.

Salem, R.M., and Burdette, E.G., 1998. Role of chemical and mineral admixtures on physical properties and frost –resistance of recycled aggregate concrete. *ACI Materials Journal*, Vol. 95, No. 5, pp. 558–563

"The building commissioning guide," U.S. General Services Administration Public Buildings Service Office of the Chief Architect, April 2005.

Tricker, R., "ISO 9000 for small business," Butterworth-Heinemann, 1997. Project Management Institute Standards Committee. A Guide to the Project Management Body of Knowledge. Upper Darby, PA: Project Management Institute, 2004, 1997.

Van Acker, A., Recycling of concrete at a precast concrete plant. FIP Notes, Part 1: No 2, 1997, pp 3–6; Part 2:No.4, 1997, pp. 4–6, 1997.

Yuan, R.L., et al., 1991. Evaluation of core strength in high-strength concrete, *Concrete International*, Vol. 13, No.5, pp. 30–4.

5 Loads and Structure Failure

5.1 INTRODUCTION

One of the main reasons for structure failure is the load. However, as discussed before, in most cases the failure occurs due to the following two factors: low structural strength and high loads. Really, the event of high load can be measured and predicted rather than the weak of the strength. High loads may cause due to change in the use of the building occupancy or due to a change of using the structure in general. For industrial structures, it controls by management of change procedure that was followed in most international oil and gas companies and other big industries as it is part of ISO 9001.

Really the problem is in the residential buildings, offices or other medium or small industrials that they do not have a management of change procedure.

As a case study of that one of the international hotel in Cairo they change the first use of one of the room in the first floor by putting a clay and start to put some plant. So the clay thickness is about 150 mm with density around 1.6 t/m^3 so can you imagine how the value of the increasing in dead load. This cause a destroy of the room floor slab which damage around three cars in the parking garage underneath this floor.

So, let us discuss and understand clearly together the characteristics of load that is used in our design to let us feel how our structure behavior under the load.

One of the major uncertainties that can affect the structure is the load. All the loads that affect the structure are considered variable as the dead load; however, the live load is variable with time and there are many researches and tests that have applied to define the variability on the live load. Therefore, this chapter will try to present all of the important researches with different approaches to define the phenomena of the live load and its characteristics.

To define the load probability distribution that affects the building during its life time will be presented in this chapter. This chapter will discuss all the loads that affect the building and how one can obtain its statistical values and use probability distribution to obtain the structure reliability.

Recently, by using computer software in structural analysis, the software calculated the self weight of the structure member after defining its materials and dimensions. However, the finishing materials and the dead load that occupied the building should be put manually in the program. Hence, it is not just a number to put in the computer, it is feeling of the effect of uncertainty of this value on the structure reliability. So, in the design process, the designer should define exactly the dead load from the concrete self weight tiles, the type of materials that will be used in finishing the roof, and other loads. After his great efforts, there still exists a variation in these values that will be covered by the safety factor for the loads in designing the concrete structures. Therefore, it is required to clearly understand the variation accompanying in these loads and the source of these variations.

5.2 DEAD LOAD

For the dead load effect, we need to differentiate between error and deviation or tolerance as the first is prohibited and the second is covered by the code and included inside the safety factor as we discussed before.

The total permanent load that must be supported by a structure is generally the sum of the self-weights of any individual structural elements and other parts. The uncertainty of a self-weight load effect comes, as pointed out by Ditlevsen (1988a), from two sources: variability of the unit volume weight and the variability of geometrical measures. Ditlevsen also pointed out that permanent loads are well represented by the normal probability distribution.

Thoft-Christensen and Baker (1982) concluded that when the total permanent load acting on a structure is the sum of many independent components, the coefficient of variation of the total load is generally much less than those of its components.

According to Williams (1995), dead load can be modeled as either normally distributed (Rosowsky and Ellingwood, 1990, 1991) or lognormally distributed (Thurmond et al., 1984, 1986). For small variation, the difference between the normal distribution and the lognormal distribution is negligible, but the lognormal distribution can guarantee positive values of dead load.

Ellingwood and Ang. (1974) estimated the variation in dead load due to structural elements to be 0.064 and variation due to non-structural elements to be 0.10, resulting in a total variation equal to 0.12.

Galambos (1980) proposed a normal probability model for the dead load with a ratio of mean to nominal values equal to 1.05, to account for the tendency of some designers to underestimate the total dead load. The coefficient of variation of dead load was estimated to be 0.10 regardless of the type of construction material. This was a part of a research work intended to evaluate the load factors for the ACI code. Galambos (1981), Galambos et al. (1982), and Mirza (1987) used the same model of dead load in their studies on the reliability of structural members.

5.3 LIVE LOAD CHARACTERISTICS

In the last decade, there have been significant advances of the application of structure reliability theory. One area of attention has been live loads in buildings, including the development of realistic stochastic models.

Rapid advances of probabilistic modeling of live loads in recent years can be attributed to a growing awareness of the designer's uncertainty about the loads acting on a structure and the acceptance of a probabilistic assessment of these loads.

The total live load is composed of two parts which are the sustained load and the extraordinary load. The summation of these two types of load is presented to obtain the probabilistic model of the total live load on residential buildings along the lifetime of the structure.

Loads and Structure Failure

The time behavior of the live load model on a given floor area can, in general, be decomposed into two parts: a sustained load and an extraordinary (or transient) load. The sustained load includes the furnishings and personnel normally found in buildings and is usually measured in live load surveys. This load is assumed to be a spatially varying random function and constant with time until a load change takes place. These load changes are assumed to occur as Poisson arrivals as shown in Figure 5.1a. The extraordinary load is usually associated with special events that lead to high concentrations of people, and it may also be due to the stacking of furniture or other items. The extraordinary load occurs essentially instantaneously and it is assumed to arrive as Poisson events. Each event is modeled by a random number of randomly load values as shown in Figure 5.1b.

The total live load history shown in Figure 5.1c is the sum of sustained and extraordinary load components, and its maximum value represents the largest total load that may be occur on a given floor area during the structural life time.

5.3.1 PREVIOUS WORK ON LIVE LOAD

Corotis and Doshi (1977) provide a brief background on the major surveys that they are analyzed in their study. The major surveys presented by Corotis and Doshi are the 1952 survey by Dunham, Brekke, and Thompson (1952) for the public building administration, the 1968 survey in Washington by Bryson and Gross for the National

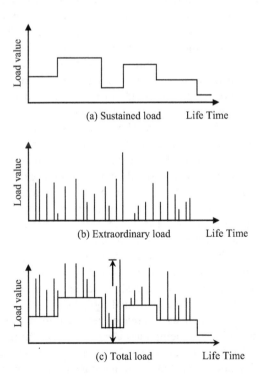

FIGURE 5.1 Live load characteristics.

Bureau of Standards (NBS), the 1969 survey in Hungary by Karman for the international council of building research, the 1971 survey in London, England, by Mitchell and Woodgate (1971) for the building research station, and the (1976) survey by Culver (1976, 1975) and McCabe et al. (1975) of the NBS.

Dunham, Brekke, and Thompson (1952) presented information on floor load surveys in two department stores, eight buildings occupied for industrial use, and two warehouses. Due to the size of the department stores, the surveys were limited to the selling areas and the small storage areas frequented by the clerks. For the buildings classified under industrial occupancy, only those buildings or parts of buildings in which some operation concerned with the primary industry were surveyed. Plans of the various buildings were presented, along with a table of total areas and area surveyed. Histograms with ranges of 240 N/m^2 and areas, in square feet, were reported, and in the case of the department stores, these were given both as surveyed and with all aisle spaces crowded.

Bryson and Gross (1968) presented survey results of two office buildings. The NBS administration building was surveyed completely, expect for the basements and the utility rooms. Out of a total of 335 rooms (6,580 m^2) surveyed, 252 rooms (4,988 m^2) were offices. In the U.S. Civil Service Commission Building, all 556 rooms (11,700 m^2) were surveyed, and 453 were offices. The live loads in Bryson and Gross' survey included the occupants, the floor covering, movable partitions, and all furniture and their contents except built-in items. The survey data included the floor level, room number, room use, overall dimensions, and number and sex of assigned personnel. The weight, location, base area, and description, along with the description and measurements of the contents, were provided. The location of each of the items had been simplified by assigning it to one of nine sections within the room, determined by intersecting column and middle strips. The service finish and the weight of trim and floor covering were noted. The weighting was made using electronic platforms. For items that could not be weighted due to size, weight, or sensitivity, or because they were secured to the floor, they were either estimated or obtained from the manufacturer's data. The survey results are provided in the form of tables and graphs, including frequency tables and histograms.

Karman (1969) presented results of a survey of 183 domestic dwellings, three office buildings, one hospital, one hospital laboratory, one health service clinic, and two schools in Hungary. Exact measurements of the positions and weights of all the furniture and equipment were made, and the data were marked onto floor plans. It is not clear whether partitions were included. Conventional bathroom scales were utilized for weight under (100 kg), and the hydraulic measuring jacks were used for heavier loads. The number of persons regularly using the area was noted. The survey included the whole floor area of a designated building except that occupied for bathrooms and toilets. Karman investigated the variations of floor load with time by considering rearrangement of all furniture and equipment every 2–3 years using histograms from the surveys for each use. In related research, Johnson (1953) investigated the occasional floor loads due to temporary accumulation of furniture and personnel. Johnson evaluated the data on 219 domestic dwellings from questionnaires filled in on the basis of estimation. From Johnson's investigations, Karman incorporated the phenomena of the extraordinary loads with variation of floor loads over time.

Loads and Structure Failure 85

Mitchell and Woodgate (1971) surveyed 32 office buildings in London, England, chosen randomly from a list of those built between 1951 and 1961. The detailed dimensional plan with positions of all furniture and equipment was constructed. The normal number of personnel was determined by observation, and questioning to determine maximum past crowding was included. Simultaneously, a schedule was prepared giving the description and weight of each item. Weights were obtained by directly weighting of items using a specially designed apparatus, by bulk density, from a manufacturer's list, or by estimation based on experience. The weights of personnel were obtained by conducting a separate survey. Partitions were omitted. Mitchell and Woodgate investigated the vertical column "stacking" effect, which is the effect on the reliability of heavily loaded bays located above each other on successive floors. They also investigated the variation of loading with time due to change in occupancy, noting that the average period of occupancy is 8.8 years. A study of the effects of load concentration was also made for the various bay sizes with free, simply supported, and fixed edges, using Levy's (1959) method of analysis.

The NBS completed a survey of fire loads and live loads in buildings in the United States. Culver and Kushner (1975) published the first a three-report series explaining the philosophy and methodology being used in the survey. The second report by McCabe et al. (1975) described the data collection and data processing procedures utilized in connection with the NBS survey. Twenty-three private and government office buildings throughout the United States were selected to identify the important factors potentially significant to the survey. The areas surveyed within the buildings were restricted to offices and related work areas, and load magnitude was determined for a single point in time, although the buildings surveyed were to monitored over time in subsequent follow-up surveys with the object to determining the loading history.

From the previous described surveys, Corotis and Doshi (1977) obtained the histograms and basic statistics of results. Corotis and Doshi fit the data to the three common probability distributions: the normal, lognormal, and gamma distribution by the method of moments. Finally, the observed data and models were plotted and quantitative tests on the goodness-of-fit conducted. They found from visual inspection of plotted histogram and cumulative graphs and from quantitative goodness-of-fit tests that the gamma distribution adequately describes the room-wide sustained live load. This distribution seems to provide a good estimate of the observed loads in the upper tail region and is more consistent than both the normal and lognormal distributions. The normal model generally underestimates the observed values in the upper region.

5.3.1.1 Statistical Model for Floor Live Loads

There are different models used to represent the floor live load. Hasofer (1968) suggested that the tail of live load distribution can be well approximated by Pareto formula ($c \cdot X^{-\alpha}$). Hasofer obtained the values of the equation parameters by analysis of the survey data.

The floor live load consists of both sustained and extraordinary load process. Murphy et al. (1987) found that the failure usually occurs when the sustained load and extraordinary load both act on the structure. Therefore, more researches studied the effects of the combined live load.

The Type-I extreme-value distribution was used to model the 50-year maximum floor live load. The occurrence of combined live load was assumed to be once a year with a 1-week duration (Rosowky and Ellingwood, 1990, 1991).

From the available load surveys, Corotis and Chalk (1980) developed a probabilistic format for the determination of building design, floor live loads by examining in detail the collection of live load data, and the behavior of the live load process. Corotis and Chalk found that for the single-tenant or one-floor case, the Type-I model of the maximum sustained load over the design life time of the building gives accurate results of a wide probability range. For multiple tenants or multiple floors, the life time maximum sustained load is described well by the largest of a Gauss–Markov sequence of correlated load changes.

Cornell and Pier (1973) proposed a probabilistic model of sustained floor loads applied to office buildings. This model encompasses and generalizes most previous reported models, and the application employs data from the most significant load survey to date 1972. Cornell and Pier proposed a linear probabilistic load model to represent the load intensity at a point in the building at an arbitrary point in time. They found, also, that the load intensity model is virtually independent of structural type.

Wen (1977) derived a probability distribution of the maximum combined load effect over a given time interval. He concluded that risk due to simultaneous occurrence of extreme loads can be evaluated from the mean occurrence durations, and intensity distributions of individual loads. When the distributions of the load under consideration are of similar characteristics, the maximum combined load effect may be described approximately by one of the asymptotic extreme-value distribution. The parameters of the distribution can be derived analytically from the first two moments and occurrence rates of individual load.

Harris et al. (1981) developed models of building live loads to include a more realistic representation of extraordinary loading situation and a detailed study of load with contributory area. They found that several simplifications in computing life time maximum total loads were shown by simulation to be reasonable. A multiple extraordinary load model was introduced to more realistically model the physical loading process.

Jaria and Corotis (1979), by studying the live load, found that each extraordinary load is assumed to be a gamma distribution and independent, and the simultaneous occurrences are assumed to be Gaussian.

Ellingwood and Culver (1977) analyzed the data of the first load survey for office buildings conducted by the NBS on 2,226 rooms in 23 buildings. They assumed an extreme-value Type-I probability model for the total maximum live load.

Using a survey data completed in 1979 on a total of 11 buildings in Sydney, Choi (1990) introduced an equation to evaluate the 955 fractile live load (characteristic live load) as a function of the influence area. He recommended that the live load in the codes should be a function of the influence area by applying a reduction factor to the live load for larger influence areas.

Corotos et al. (1981) used a Delphi method to discuss applicable live load design values. By using this method, they reduced the nominal live load for the office

Loads and Structure Failure

corridors above the first floor, residential corridors in hotel which do not serve public rooms, and a residential corridors in multifamily buildings.

5.3.2 STOCHASTIC LIVE LOAD MODELS

The problem of stochastic load combinations has been of interest for some time as most loads acting on structures are of random nature, in addition of many such loads fluctuating with time. In assessing the safety of a structure under combinations of such loads, the problem then arises of evaluating the peak combined load during an anticipated useful life of the structure.

The nature of this problem clearly requires solutions within the scope of the theory of random process, where loads are described as stochastic processes and the life time peak load or load effect is obtained as the extreme of a function of stochastic processes over a specific period of time.

Armean et al. (1978) studied the second moment combination of stochastic load. A model that has often been used is the stationary Gaussian process. It is well known according to Parzen (1967) that a linear combination of such process is itself Gaussian. Based on this, the distributions of the extreme of linearly combined Gaussian process are evaluated by Bolotin (1969). A vector process approach had been presented by Cornell et al. (1977) for nonlinear combination of Gaussian process. The obtained distribution of the extreme in this case is only good at its far tail.

Other stochastic load models that have been studied (Borges and Castanheta, 1972; Gumbel, 1958) include the sequence of mutually independent and identically distributed random variable. The Poisson sequence wave process and the filtered Poisson process with and without finite duration at each occurrence have been studied by Bosshard (1975), Grigoriu (1975), MacGuire and Cornell (1974), and Wen (1977a, b). These load models are described in detail in Chapter 3.

Corotis (1979) studied the stochastic nature of building live loads to analyze the physical aspects of live load behavior in order to develop a model that will form the basis of future surveys and code specification. In this study, they found that live load may be considered primarily by a function of a room use and room size as in recent survey by Culver (1976). Culver and Ellingwood (1977) showed that there was a tendency of the load to decrease when the area of the room increases. In fact, Culver concluded that the mean of the live load decreases (for unspecified use) 0.00134 psf for each square foot of area (0.691 Pa/m^2). He also noted that this rate changed when the use is specified.

Corotis and Tsy (1983) used the stochastic live load model, including both sustained and extraordinary loads, as the basis of deriving load duration statistics. The sustained load, which includes the furnishings and personnel normally found in buildings, was assumed to be Gamma distribution, and the time between changes of the sustained load was assumed to be exponentially distributed. The extraordinary load which is usually associated with special events that lead to high concentrations of people was assumed to be gamma distributed and its occurrence to be Poisson distributed.

Corotis and Doshi (1977) presented the sustained load by a gamma probability distribution and the duration by the exponential distribution, while the probability density function (PDF) of a single extraordinary event was assumed to be gamma, and the maximum extraordinary load during each sustained load duration was generated directly from a Type-I extreme-value distribution.

Most live loads acting on structures are of random nature. In addition, many such loads fluctuate with time. Determining the total combination of such loads requires a solution within the scope of the theory of random processes wherein loads are described as stochastic processes.

According to Renold (1989), the definition of a stochastic process is a collection of random variables when the collection is infinite. The collection will be indexed in some way, e.g., by calling {x(t):t∈T} a stochastic process, we mean that x(t) is a random variable for each (t) belonging to some index set (T) (more generally, x(t) may be a vector of random variables).

Usually the index is either an interval T = [0,∞], or a countable set such as the set of nonnegative integers t = {0, 1,...}, and t will denote time with these alternatives of T. We have either a continuous time or discrete time process, respectively.

Starting with Borges and Castan Heta's work (1972), a number of models for describing the variability of loads have been suggested by Bosshard (1975), Grigoriu (1975), Larrabee (1978), Pier and Cornell (1973), and Wen (1977). Among these models, the most general yet tractable models are the Poisson square wave (PSW) and the filter Poisson (FP) processes.

5.3.2.1 Poisson Square Wave Process

This process has been used to model sustained loads. It is assumed in this model that the load intensity changes at random points in time following a Poisson process. Intensities within these points are assumed to be independent and identically distributed random variables.

If (v_s) denotes the mean rate of load changes, and $F_s(s)$ describes the common cumulative density function (CDF) of intensities, Parzen (1967) shows that the occurrences of S > s following load changes constitute a Poisson process with mean occurrence $v_s [1-F_s(s)]$. The CDF of the extreme may, therefore, be evaluated as follows:

$$F_{S_T} = P(S_T < s) = P(S < s)_{t=0} P(noS > s \text{ following load changes})_{0 \to t}$$
$$= F_S(s)\exp\{-v_s T[1-F_s(s)]\} \tag{5.1}$$

Upon differentiating, the PDF is obtained as follows

$$f_{S_T}(s) = f_S(s)[1+v_s TF_s(s)]\exp\{-v_s T[1-F_S(s)]\} \tag{5.2}$$

where $f_s(s)$ is the PDF of S and T is the total lifetime.

5.3.3 Filtered Poisson Process

This process has been used to model extraordinary loads with random occurrences. It is assumed in this model that load occurrence follows a Poisson process, and that

Loads and Structure Failure

load intensities at various occurrences are independent and identically distributed random variables. The process is generalized to include random load duration, denoted by t, which is also assumed to be statically independent and identically distributed at various occurrences.

Let v_r denotes the mean occurrence rate of the extraordinary load, $F_R(r)$ represents the distribution at each occurrence, and let N be the random number of occurrence during T. On the basis of the total probability theorem, the CDF of R_T is obtained as follows:

$$F_{RT} = P(R_T < r) = \sum_{n=0}^{\infty} P(R < r/N = n)P(N = n) \tag{5.3}$$

$$\text{in which } P(N = n) = \frac{(v_r T)^n \text{Exp}(-v_r T)}{n!} \tag{5.4}$$

and

$$\begin{aligned} P(R < r/N = n) &= [F_r(r)]^n && \text{for } n = 1,2,3,\ldots \\ P(R < r/N = n) &= 0.0 && R < 0.0; \text{ for } n = 0 \\ P(R < r/N = n) &= 1.0 && R > 0.0; \text{ for } n = 0 \end{aligned} \tag{5.5}$$

The last equation neglects the effect of overlapping of occurrences and is, therefore, acceptable when overlapping is unlikely or when load intensities at overlaps do not accumulate. Substituting Equations 5.4 and 5.5 with Equation 5.3, one can obtain the following equations:

$$F_{RT}(r) = \sum_{n=0}^{\infty} [F_R(r)]^n \frac{(v_r T)^n \text{Exp}(-v_r T)}{n!}$$

$$F_{RT}(r) = [F_R(r)]^0 \text{Exp}(-v_r T) + \sum_{n=1}^{\infty} [F_R(r)]^n \frac{(v_r T)^n \text{Exp}(-v_r T)}{n!}$$

$$= [F_R(r)]^0 \text{Exp}(-v_r T) + F_R(r) v_r T \cdot \text{Exp}(-v_r T) + [F_R(r)]^2 \frac{(v_r T)^2}{2!} \text{Exp}(-v_r T) + \cdots$$

$$F_{RT}(r) = \text{Exp}\{-v_r T[1 - F_R(r)]\} + (H_r(r) - 1)\text{Exp}(-v_r T) \tag{5.6}$$

where H(r) is unit step function

$$\begin{aligned} H_r(r) &= 0 && \text{if } r < 0.0 \\ H_r(r) &= 1.0 && \text{if } r > 0.0 \end{aligned} \tag{5.7}$$

Upon differentiating Equation 3.6, the PDF of R_T is obtained as

$$f_{R_T} = v_r T f_R(r) \exp[-v_r T[1 - f_R(r)]] + \delta(r) \exp(-v_r T) \quad (5.8)$$

where $\delta(r)$ is the Dirac Delta function representing a spike unit area at r = 0.0.

5.3.4 ANALYSIS OF THE SUGGESTED MODEL

The total live load is the summation of sustained load and extraordinary load over the lifetime of the structure.

To describe the sum of the sustained and extraordinary random load variables, consider the independent continuous random variables, s and r, with PDF $f_s(s)$, $f_R(r)$ and cumulative distribution functions $F_s(s)$, $F_R(r)$, for sustained and extraordinary load, respectively. Assuming independence, the joint PDF $f_{S,R}(s,r) = f_s(s) \cdot f_R(r)$ (Cornell and Benjamin, 1974). It is required to determine the PDF of the total load, l, where l is linear function of s and r:

$$l = s + r$$

We first find CDF of l where the joint PDF of s and r is integrated over the region $s + r \le l$ as indicated in Figure 5.2:

$$F_L(l) = P(L \le l) = P(l \le s + r) \quad (5.9)$$

$$= \int_{-\infty}^{\infty} \int_{-\infty}^{l-r} f_{S,R}(s,r) \, ds \, dr$$

$$= \int_{-\infty}^{\infty} \int_{-\infty}^{l-r} f_S(s) f_R(r) \, ds \, dr$$

$$= \int_{-\infty}^{\infty} F_S(l-r) f_R(r) \, dr$$

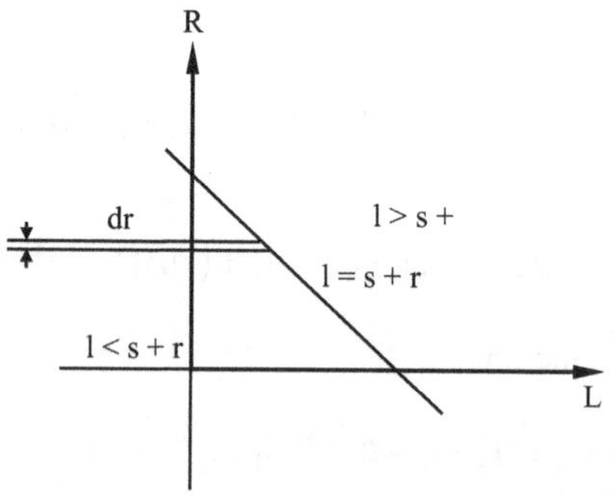

FIGURE 5.2 Region of summation in the case of two random variables not restricted to nonnegative values.

Loads and Structure Failure

Then, the corresponding density function (PDF) can be obtained by differentiating the CDF of Z as follows:

$$f_L(l) = \frac{\partial F_L(l)}{\partial L}$$

$$= \frac{\partial}{\partial L} \int_{-\infty}^{1-r} F_S(1-r) f_R(r) \, dr$$

$$= \int_{-\infty}^{\infty} f_S(1-r) f_R(r) \, dr \quad (5.10)$$

where the density f_L is called the convolution of probability densities $f_S(s)$ and $f_R(r)$. For nonnegative random variables as in the case of live loads, the range of integration in Equations 5.9 and 5.10 shown in Figure 5.2 reduces to the region indicated in Figure 5.3 because $f_S(s)$ and $f_R(r)$ are zero for negative argument. Therefore, Equations 5.9 and 5.10 can be written as follows:

$$F_L(l) = \int_0^1 F_S(1-r) f_R(r) \, dr \quad (5.11)$$

$$f_L(l) = \int_0^1 f_S(1-r) f_R(r) \, dr \quad (5.12)$$

Substituting Equations 5.1, 5.2, 5.6, and 5.8, for the PDF and the CDF for sustained and extraordinary loads, respectively, with Equations 5.11 and 5.12, one can obtain the PDF and CDF of the total load as follows:

$$f_L(l) = \int_0^1 f_s(s)[1+v_s T f_s(s)] \operatorname{Exp}\{-v_s T[1-f_s(s)]\} v_r T f_R(1-s) \operatorname{Exp}\{-v_s T[1-f_s(1-s)]\} \, ds$$

$$(5.13)$$

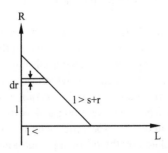

FIGURE 5.3 Region of summation in the case of two random variables restricted to nonnegative values.

$$F_L(l) = \int_0^1 \text{Exp}\{-v_r T[1 - F_R(l-s)]\} f_s(s)[1 + v_s T F_s(s)] \text{Exp}\{-v_r T[1 - F_R(s)]\} ds$$

(5.14)

where
T is the life time of the structure,
v_s is the mean rate of load change in case of sustained load,
v_r is the mean occurrence rate of the extraordinary,
$F_s(s)$ and $f_s(s)$ are the CDF and the PDF of the sustained load, respectively, and
$F_R(r)$ and $f_R(r)$ are the CDF and the PDF of the extraordinary load, respectively.

Knowing the structure life time, the duration of the sustained load, the rate of occurrence of extraordinary load, and the PDF and CDF of the sustained and extraordinary loads and performing the integration in Equations 5.13 and 5.14, one can find the values of the PDF; $f_L(l)$, and CDF; $F_L(l)$, corresponding to the value of the total load l.

Several researchers discussed these parameters and their value for different types of building occupation.

Chalk and Corotis (1980) summarized the most survey data results for sustained and extraordinary load. They found that the gamma distribution presents the sustained load value and the exponential distribution is used to represent the time between changes. On the other hand, the extraordinary load value is presented by a gamma distribution and its occurrence to be Poisson distribution.

After reviewing of survey data, Corotis and Doshi (1977) substantiated the use of gamma probability distribution of the magnitude of sustained load. On the other hand, Cornell and McGuire (1974) suggested to use gamma probability distribution to present the PDF of a single extraordinary event.

In ANSI code, the statistical parameters for sustained and extraordinary load and the duration of sustained load and rate of occurrence of extraordinary load are presented for different occupation types as shown in Table 5.1. Moreover, the reference periods of different occupancies are presented.

According to Michael and Corotis (1981), all extraordinary live loads occupancy duration are assumed to follow a uniform distribution with mean value of 2 weeks, 6h, and 15 min. Kiureghiam (1980) in his study takes the duration of the extraordinary load equal to 3×10^{-3} year.

According to Kiureghian (1978, 1980), the extraordinary load durations are short enough with respect to mean sustained load durations in that the extraordinary loads may still be considered a point process when combining with the sustained loads for life time total statistics. The sustained load which occurrence during the structure life time (T) only is of interest considering the sustained load interval of PSW process between zero and the first transition point and noting that a portion of the last interval is generally truncated. Therefore, the mean rate of sustained load duration will be modified and calculated by the following equations:

$$v_{ok} \cong v_r v_s (\tau_r + \tau_s)$$

Loads and Structure Failure

TABLE 5.1
ANSI Code (A58.1-1982) Typical Live Load Statistics

Occupancy	Sustained Load m_s (kg/m²)	σ_s (kg/m²)	Extraordinary Load m_r (kg/m²)	σ_r (kg/m²)	Temporal Constants τ_s (years)	ν_e (per year)	T (years)
Office Buildings							
Offices	53.2	28.8	39.1	40.0	8	1	50
Residential							
Owner occupied	29.3	12.7	29.3	32.2	2	1	50
Renter occupied	29.3	12.7	29.3	32.2	10	1	50
Hotels							
Guest rooms	22.0	5.9	29.3	28.3	5	20	50
Schools							
Class rooms	58.6	13.2	33.7	16.6	1	1	100

$$v_{sm} \cong \left(\frac{0.5}{T} + v_s\right)\frac{v_{ok}}{v_s + v_{ok}} \tag{5.15}$$

where v_{ok} is the mean rate of coincidence in this combination,
τ_r is the duration of the extraordinary load,
τ_s is the duration of the sustained load, and
v_{sm} is the modified sustained load, mean rate.

Table 5.1, from American National Standard Institute (ANSI), presents the average values for live loads and standard deviation for different buildings for sustain and extraordinary load.

5.3.5 Methodology and Calculation Procedure

The integration of Equations 5.13 and 5.14 by the analytical procedure is very complicated and time-consuming. The method of Romberge numerical integration technique which is described in many mathematical books such as Grahan (1986) is used in this section to obtain the value of the total live load.

The Romberge numerical integration technique is performed by using the Microsoft Excel software program (1996). The PDF of the total live load is calculated assuming that the parameters of Equations 5.13 and 5.14 which is used in the integration are as follows:

- The structure life time (T) is 50 years.
- The sustained live load is assumed a gamma distribution with mean value kg/m² with standard deviation 16.6 kg/m² with mean occurrence rate, v_s, is 0.5/year.

- The extraordinary load is assumed to have a Gamma distribution with mean value equal to 29.29 kg/m² and standard deviation 32.22 kg/m² and the mean occurrence rate 1.0/year.
- The duration time of extraordinary load, τ_r, is 3×10^{-3} year. Therefore, the modified sustain load mean occurrence rate is calculated from Equation 5.15 and it is found to be 0.34/year.

The PDF of total live load is obtained as plotted. The mean and standard deviation of the total load are calculated and it is found that the mean total load in case of residential building is equal to 106.13 kg/m² and the standard deviation equal to 42.52 kg/m².

5.3.6 Testing of the Suggested Model

After the model distribution is obtained by the previous integration method, the data are tested with the common distributions by using chi-square (χ^2) which is goodness-of-test is applied to continuous random variables and its value is calculated from the following equation:

$$\chi^2 = \sum_{i=1}^{k} \left[\frac{(O_i - E_i)^2}{E_i} \right] \qquad (5.16)$$

where O_i and E_i are the observed and expected number of occurrences in the ith interval, respectively, and k is the number of interval.

Moreover, the Kolmograv–Smirnov (K–S) test is done which is a second quantitative goodness-of-fit test based on a second test statistic. It concentrates on the deviations between the hypothesized cumulative distribution function and the observed cumulative data, and the calculation of the test is based on the following equation:

$$K - S = \max_{i=1}^{n} \left[\left| \frac{i}{n} - F_X(X^{(i)}) \right| \right] \qquad (5.17)$$

The results of the two tests with some hypothesized probability distribution are shown in Table 5.2 and from this table one can find that the least value for the two tests is at hypothesize a lognormal distribution. In addition, the relations between the suggested model and the different probability distributions are shown in Figure 5.4 which clearly shows that the model is more coincidence with the lognormal distributions than others.

TABLE 5.2
Results of Goodness-of-Fit Tests for Comparison between the Suggested Model and Some Common Probability Distributions

Probability Distribution	χ^2	K–S
Lognormal distribution	0.004644	0.01226
Extreme-value Type I	0.019781	0.02022
Gamma distribution	0.04796	0.036673

Loads and Structure Failure

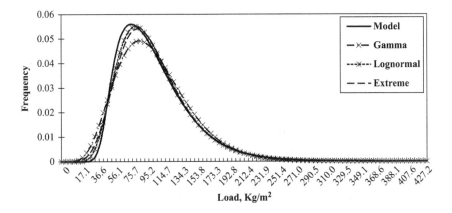

FIGURE 5.4 Comparison between the model and different probabilities distributions.

Moreover, the test by the above two methods is performed by the Crystall Ball program (1996) to test the suggested model with the more traditional probability distributions.

One can see that the model is close to the lognormal distribution with mean value equal to 106.13 kg/m² and a standard deviation equal to 42.52 kg/m².

For verification of the suggested model using Monte-Carlo simulation (MCS). The research performed by El-Reedy et al. by using MCS technique for constructing a model for residential building, revealed that lognormal distribution is the best probability distribution that can present the live load for residential buildings. The comparison between the model and different distributions is shown in Figure 5.4.

In order to verify the analytical model with the suggested load of occupancy, an MCS technique (Ang and Tang, 1984) is used to simulate the total load process by using the so-called Crystall Ball program (1996) as follows:

1. Generate the magnitude of the sustained load from the gamma distribution.
2. Generate the duration of the sustained load at the previous magnitude from the exponential distribution.
3. From the previous duration time and by using Poisson distribution, generate the number of the extraordinary load that occurs during the sustained load duration.
4. Generate the extraordinary load magnitude for every load from extraordinary load gamma distribution.
5. Calculate the total load value by summing the sustained load and every extraordinary load.
6. Repeat steps 1–5.

The MCS is performed for 10,000 trials. The values from the trials are divided into 5 kg/m² intervals. The frequency of occurrences in each interval is plotted as a bar and the corresponding values from the suggested model as a curve as shown in Figure 5.5. From this figure, one can find that there is no difference between the suggested analytical load model and MCS results.

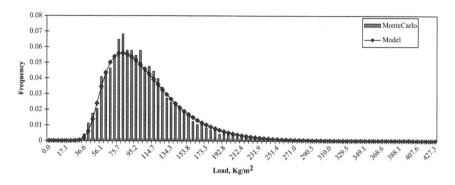

FIGURE 5.5 Comparison between the model and Monte-Carlo simulation results.

The analysis given above shows that the live load effect along the life time of the structure in the case of residential building taking into consideration the life time of the structure years can be presented by a lognormal distribution with mean value equal to 106.13 kg/m² and standard deviation equal to 42.52 kg/m².

Rapid advances of probabilistic modeling of live loads in recent years can be attributed to a growing awareness of the designer's uncertainty about the loads acting on a structure and the acceptance of a probabilistic assessment of these loads.

As a normal procedure, from the probability distribution, one can obtain the definite number that can be used easily by the design engineer as the equations of design for any code are based on the deterministic analysis. But the risk and probability of failure for any code to another is different for the values of the loads.

5.4 LIVE LOADS IN DIFFERENT CODES

The values of live load and its reduction factors are different from onecode to other, depending on the social life of the country that is applying this code.

Moreover, the resistance reduction factor is different from one code to other, depending on the quality control, capabilities of the contractors, and the supervision government system of each country.

Therefore, the reliability of reinforced concrete structure is different from one code to other depending on the different factors of safety of load and resistance.

In the following section, a comparison between live load values for different codes (ECP, BS8110, ANSI) is presented.

5.4.1 COMPARISON BETWEEN LIVE LOAD FOR DIFFERENT CODES

For the limit state design method, different codes consider live load factor as this factor accounts the unavoidable deviations of the actual load from the code value and for uncertainties in the analysis that transforms the load into a load effect.

The load reduction factor is recommended by different codes to account for the decrease in of total load due to less probability of applying the same value of load on different floors in the same time.

Loads and Structure Failure

The values and factor of live load are different from one code to other, while the reduction factor of live load is different from one floor to another in residential buildings for different design codes.

Therefore, in the following sections, comparisons between design codes are presented to show the difference between the values of live load, limit state load factors, and reduction of floor load in their codes. Codes which are considered in this comparison are as follows:

- Egyptian code of practice (ECP).
- British code (BS8110).
- American National Standard Institute code (ANSI).

5.4.2 Values of Live Loads and Its Factors in Different Codes

The live load values taken into consideration in design are different from one code to another, depending on the variation on the social life from one country to other.

The different live load values and the load factor in a limit state design method for different codes are shown in Table 5.3.

From this table, one can find that there are different values of live load and load factor for the presented codes.

The Egyptian code is more conservative for the value of live load. However, the factor of live load is the same in the limit state equations for most of codes, but ANSI code is conservative than others.

Now it is very important to illustrate the methods of calculating live load in Egyptian, American code (ANSI), and the British code (BS8110).

The live load values considered in design will be according to building use if it is residential, administration, theater, or others. All the values for the different codes are illustrated in Table 5.4.

5.4.3 Floor Load Reduction Factor in Different Codes

The reduction factors of codes of live loads on different floors in residential buildings are different and are considered in this section (ECP, BS8110, and ANSI codes). These reduction factors for live load are summarized in Table 5.5.

TABLE 5.3
Comparison between Values and Factors Live Load in Different Codes

Code	Load value[a] (kg/m^2)	Equation
ECP	200	1.4 D.L + 1.6 L.L
BS8110	153	1.4 D.L + 1.6 L.L
ANSI	195	1.4 D.L + 1.7 L.L

[a] These values for residential buildings.

TABLE 5.4
Live Load Values in Different Codes

| | | | Live Load | |
| | | | | BS8110 |
Type of the Building	ECP (kg/m²)	ANSI (kg/m²)	Distributed Load (kg/m²)	Concentrated Load (kg)
Roof				
Horizontal surface cannot reach (no use)	100			
Inclined >20° (no use)	50			
Horizontal or inclined (can be reached and used) in residential buildings	200			
Horizontal can reach in public buildings	300			
Residential Building				
Rooms	200	195.3	153	143
Stairs	300	195.3	306	459
Balcony	300	195.3	153	153/m'
Administration Building	300			
Halls		488	408	459
Offices		244	255	275
Theaters, Cinema, and Library		500	488	408
Passenger Room without Fixed Chairs		600		408
Warehouse	1000			
Light		610.3	408/m height storage	918
Heavy		1220.61	765	459
School				
Classes		195.3	306	275
Corridors		390.6	408	459
Stairs and exits		488	408	459
Hospital				
Operation room and laboratory		293	204	459
Diagnosis room		195.3		
Reception hall		195.3	204	459
Corridor above first floor		390.6	408	459

It is worth to mention that the reduction factor in ANSI code depends on the influence area. Members having an influence area of 37.2 m² (400 ft²) or more may be designed for a reduced live load determined by applying the following equation:

$$L = Lo\left(0.25 + \frac{15}{\sqrt{A_I}}\right) \quad (5.18)$$

TABLE 5.5
Floor Reduction Factors for Live Load in Different Codes

	Floor Reduction Factor[a]		
Floor	ECP	BS8110	ANSI[a]
Roof	1.0	1.0	0.82
1st Floor	1.0	0.9	0.65
2nd Floor	0.9	0.8	0.58
3th Floor	0.8	0.7	0.54
4th Floor	0.7	0.6	0.51
5th Floor	0.6	0.6	0.48
6th Floor	0.5	0.6	0.47
7th Floor	0.5	0.6	0.45
8th Floor	0.5	0.6	0.44
9th Floor	0.5	0.6	0.43
10th Floor	0.5	0.6	0.42
11th Floor	0.5	0.5	0.41
12th Floor	0.5	0.5	0.41
13th Floor	0.5	0.5	0.41
14th Floor	0.5	0.5	0.41
15th Floor	0.5	0.5	0.41
"	"	"	"
"	"	"	"

[a] Calculated from Equation 5.18 in case of four bay with $16\,m^2$.

where L is the reduced design life load per square foot of area supported by the member,

Lo is the unreduced design live load per square foot of area supported by the member, and

A_I is the influence area, in square feet.

The influence area A_I is four times the tributary area for a column. For instance, in the case of interior column, the influence area is the total area of the floor surrounding bays.

The total load carried by columns at a certain floor level is calculated by summing the load reduction factors, as given in Table 5.5, for the floors above this level as a function of load values given in Table 5.4 for the considered codes. The results are given in Table 5.5 as a function of P_1, P_2, and P_3 which are different live load values in ECP, BS, and ANSI.

Moreover, the total reduction of live load carried by columns at different floors varies from the ECP to that of BS8110 and ANSI codes. These values in ECP are greater than that in the BS8110 and ANSI. However, these values are the same in the ECP and BS8110 from the 10th to the 20th floors.

TABLE 5.6
Comparison between Reduction Factor in Different Codes

No. of Floors	ECP	BS8110	ANSI
6	$5P_1$	$4.6P_2$	$3.58P_3$
8	$6P_1$	$5.8P_2$	$4.5P_3$
10	$7P_1$	$7P_2$	$5.4P_3$
12	$8P_1$	$8P_2$	$6.2P_3$
14	$9P_1$	$9P_2$	$7P_3$
16	$10P_1$	$10P_2$	$7.8P_3$
18	$11P_1$	$11P_2$	$8.6P_3$
20	$12P_1$	$12P_2$	$9.4P_3$

Note: P_1, P_2, and P_3 are the different live load values given in Table 5.4.

Generally, the ANSI code provides lower values for the total live load reduction factor than other codes (Table 5.6).

5.4.4 Comparison between Total Design Live Load Values in Different Codes

After comparing the reduction factors and live load values in different codes, the total design live load of a column carrying a certain number of typical floors is compared in the different design codes. The total design live load of that column is calculated by summing the floor loads which is multiplied by the corresponding reduction factor specified by codes.

The ratio between the total live load value calculated from ECP and that from other codes is then obtained, as presented in Table 5.7.

From this table, one can find that the total live load value which is taken in design according to ECP is more conservative than the values obtained from the other codes.

It is also observed that the total live load calculated by using the ECP is higher than that obtained from BS8110 and ANSI, but the ratio between values of the total live load calculated from ECP to that value obtained from ANSI and BS8110 is different from one floor to another.

In the case of six floors, the Egyptian code is more conservative in design live load value of about 42% than the British code.

Moreover, ECP live load design value is higher than BS8110 for 10–20 floors for about 31%.

On the other hand, the ratio between the live load calculated from ECP to that from ANSI code has a minimum value equal to 1.17 at one floor only but the highest ratio is 1.35 at six floors with average values 1.25.

In summary, the Egyptian code in calculating live load in the case of a limit state design method is more conservative than other codes of design.

Loads and Structure Failure

When making a comparison between different specifications by case study for building of six floors at 20-storey, the parameters (P_1), (P_2), and (P_3) are the values of live load in the specifications.

One can find that the Egyptian code is more conservative than that of the British code and the American code. We find that these factors are affected by the habits and legacies of people. So it is very important and clear that each country's specifications and codes must stem from the country where the activities of the people and their use of the building is derived from the behavior of the people. This behavior is the product of economic, political, and social realities of that country.

For example, in the case of developing countries, the rules and laws for constraints to change the activity of the building from residential to administration or to industrial is less. So it should be more conservative in design parameters rather than the countries that are more restricted in applying the laws for changing the building activities.

Table 5.7 shows the code comparison between Egyptian and American and British according to the number of floors. Generally, from the table one can note that the Egyptian code is more conservative than the British code by about 30% and gives the highest values of the American code by about 23%. These differences are fixed for 16 floors.

5.5 DELPHI METHOD

Corotis et al. (1981) discussed the using of the Delphi method for theory and design of load applications. The Delphi method is a highly formalized method of communication that is designed to extract the maximum amount of unbiased information from a panel of experts. By using this method, the most reliable conclusions should be obtained from a panel of equal expertise members. Although the method has been used for nearly 30 years in strategic planning, the applications in civil engineering have come only much more recently. No documented use in structural engineering could be found.

TABLE 5.7
Relationship between Design Live Load for Egyptian Code and Different Codes

Floor No.	EC/BS8110	EC/ANSI
1	1.31	1.17
6	1.42	1.35
8	1.35	1.29
10	1.31	1.26
12	1.31	1.24
14	1.31	1.24
16	1.31	1.23
18	1.31	1.23
20	1.31	1.23

The Delphi method is especially well suited for such structural engineering applications as setting design loads, developing uniform terminology, and selecting appropriate levels of safety. Its success depends principally on the careful selection of the panel and formulation of the questions. Its use in updating the design live loads for the ANSI A58 Standard was judged to be appropriate and fruitful. The principal difficulties were in maintaining the high level of response and in reaching and implementing a consensus.

The selection of an appropriate panel of experts was clearly the most important step in assuring the success of the Delphi. It was decided that the panel should consist of practicing engineers with extensive design experience. In order to obtain the most valuable opinions and avoid having to weigh the responses, only engineers who were held in high esteem by their peers were selected. In order to accomplish the selection, inquiries were made of engineers in several major cities. In each case, the engineers were asked to name those structural engineers in their area of the country who were highly regarded. From the composite list, 25 names were chosen. It was hoped that this would produce at least 20 responses which was somewhat arbitrarily selected as a viable minimum.

The 25 members of the panel represent a wide geographical distribution, with nine from the East Coast, eight from the West Coast, four from the Midwest, and four from the Southeast. Fourteen of the firms represented may be classified as structural engineering, six as architecture-engineering, two as civil and structural engineering, two as material supplier associations, and one as construction. All 25 panel members are structural engineers, and more than half are principals in their firms. A list of the panel members and their firms is shown in Table 5.8 and Figure 5.6.

The selection of design loads for codes and standards involves data collection and reduction, fitting of physical and empirical models, and a high degree of engineering judgment. The Delphi method is a formalized procedure for combining responses from professional sources. It differs from a Bayesian approach in which subjective opinion and objective data are incorporated by individual outside of the formal Delphi. The characteristics, basic philosophy, and historical development of the Delphi method are presented in this chapter, along with general guidelines for applications in structural engineering. This is followed by a description of the actual use of the Delphi method for updating the (unreduced) uniformly distributed live loads specified in the American National Standard –AS8 "Building Code Requirements for Minimum Design Loads in Buildings and Other Structures."

The Delphi Method is a highly structured form of communication that can be used by a panel of experts to enable a consensus to be reached. The actual procedure of applying the steps of the Delphi method is commonly termed as Delphi.

The Delphi method has four major characteristics. These characteristics have been varied in their execution, but they remain essentially basic to all applications.

The first of these is anonymity of response: the experts taking part in a Delphi are not informed of the source of individual responses and often are not told the identity of the other experts taking part in the Delphi. A controlled feedback of these responses is the second characteristic of the Delphi method. This feedback may include individual responses or questions, depending on the design of the Delphi,

TABLE 5.8
Minimum Uniformly Distributed Live from Delphi Method

	Loads (pounds per square foot)			
Occupancy Type (1)	ANSI 1972 (2) (2)	Delphi Round One (3)	Delphi Round Two (4)	ANSI 1980 (3) (5)
Assembly halls with fixed seats	60	10 at 50 6 at 60 1 at 100	12 at 50 5 at 60 2 at 100	60
Library corridors above first floor	80	2 at 50 2 at 60 1 at 75 7 at 80 4 at 100	3 at 50 8 at 60 3 at 80 5 at 100	80
Marquees	75	1 at 50 5 at 60 8 at 75 2 delete	10 at 60 8 at 75 1 delete	75
Office corridors above first floor	80	9 at 50 1 at 75 4 at 80 3 at 100	11 at 50 1 at 60 2 at 80 5 at 100	50
Residential floors above first in Dwellings	30	7 at 30 8 at 40	1 at 30 18 at 40	30–40 (varies by use)
Residential corridors in hotels not Serving public rooms	80	9 at 40 1 at 75 4 at 80 1 at 100	13 at 40 3 at 60 2 at 80 1 at 100	40
Residential corridors in multifamily Buildings	80	8 at 40 1 at 75 4 at 80 3 at 100	12 at 40 2 at 60 2 at 80 3 at 100	40
Retail stores-first floors	100	6 at 75 1 at 80 9 at 100	7 at 75 12 at 100	100

but it almost always includes the third characteristic, which is a statistical description of the group response.

This is usually the mean response plus some indication of the range of responses. The fourth characteristic is the use of multiple iterations of questions and feedback to reach a consensus. This consensus does not necessarily imply total agreement among the experts on the actual answer or answers, but rather an agreement on the uncertainty of the answer.

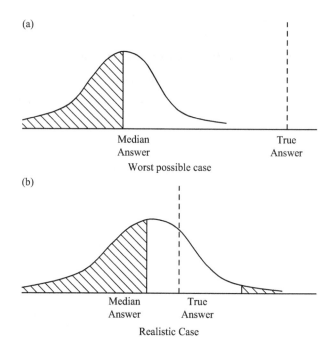

FIGURE 5.6 Example of Delphi outcomes.

A scientific investigation is termed as Lockean if its basic premise is that the empirical content of data completely determines the validity of the system as presented by Mitroff and Truff (1975). Research into the statistical nature of loads is basically Lockean, with the greatest importance assigned to the data itself and the concept that any model can only be justified by data. This Lockean philosophy that the data are separate and prior to any formal theory is true of empirical science in general and is especially true of fields that employ a statistical treatment of data.

Corridors in multifamily residences and corridors serving private rooms in hotels were reduced from 80 psf (3.8 kPa) to 40 psf (1.9 kPa) as strongly recommended in the Delphi.

Upper floors in dwellings represented a most interesting dilemma. Delphi responses were strongly applied for all habitable floors 40 psf (1.9 kPa). Supporting comments in the Delphi could be grouped into three categories. One was that waterbeds require the higher design load. The second was the 30 psf (1.4 kPa) results in too flexible a floor system. The third was that depending on the floor plan of the dwelling, heavy loads usually associated with first floors could readily be found on upper floors level or high-rise dwellings. On the other hand, no reports of inadequate performance in existing dwellings could be traced to the design live loads. A request for comments from the office of Architecture and Engineering Standards of the Department of Housing and Urban Development elicited concern over an unusual increase in housing costs that could result from requiring 40 psf. It was decided to clarify that dwelling refers only to one-family and two-family houses and to redefine the design loads to be based on room usage. Habitable attics and sleeping areas were given a design load of 30 psf (1.4 kPa), while all other habitable areas were set to 40 psf (1.9 kPa). The waterbed risk was left to the indulger.

The above recommendations were included in an April 1980 comprehensive proposed revision to the ANSI A58 Standard, and most of them were in an earlier November 1979 draft. Of those members of the ANSI A58 Standard Committee voting on the live load provisions, 84% voted for approval. The remaining votes were abstentions except for one negative, which was not related to the changes previously described.

5.6 OVERLOADS

The maximum load to come on a structure during its life will be referred to by the factor L with mean μ_L, standard deviation σ_L, and coefficient of variation V_L. The factor L can be subdivided into a number of individual factors which contribute to variations in the total load L.

The magnitudes of the loads may vary from those assumed. Although dead loads, D, are known more accurately than any other loads except possibly fluid loads, they can vary due to:

1. Variations in size of members
2. Variations in density of material due to different types of aggregate, different moisture content, etc.
3. Structural and non-structural alterations.

Thus, the deal load varies randomly from structure to structure in a population of structures and also changes from time to time in a given structure due to renovations, etc. The lifetime maximum dead load will be assumed to have an average value equal to the design value with a coefficient of variation equal to 0.07. This implies that the maximum dead load will be within ±14% of the design value in 95% of all structures.

Live load, L, varies considerably from time to time and from building to building. Extensive loading surveys suggest, however, that the occupants in a given part of a building will change from 5 to 20 times during the life of a building and hence the probability having a high load during the life of the building is fairly high. In addition, larger the area considered, the smaller the scatter in the maximum load.

The office live loads and live load reduction factors for floor members in the 1975 National Building Code of Canada agree quite closely with the 95th percentile loads from these distributions (Figure 5.7).

Allen (1975) suggested that the mean lifetime maximum office live load is about 70% of the loads specified in loading tables with a coefficient of variation of 0.3.

Different load combinations may be critical in different cases. This is especially true if the loads do not act in the same sense.

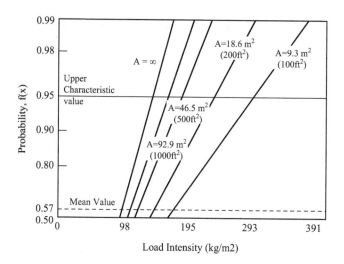

FIGURE 5.7 Distributions of maximum office floor loads expected during 50 year life of a building.

5.6.1 Uncertainties in Calculation of Load Effects

The assumptions of stiffnesses, span lengths, etc., and the inaccuracies involved in modeling three-dimensional structures for structural analysis lead to variations between the stress resultants which actually occur in a building and those estimated in the designer's analysis. Allen (17) also suggested that the mean values from an analysis should be about equal to 1.0 times the real values. For statically determinate members, Allen suggests $V_E = 0.07$. For more complex structures, Lind (15) suggests $VE = 0.10$ for beams in which the moments normally must resist the loads applied to the beams and $VE = 0.20$ for slender columns in which the moments are largely due to compatibility of deformations and hence depend on more variables than the beam moments.

Galambos and Ravindra (20) have further subdivided E into one term dealing with the structural analysis itself which is assumed to have a mean of 1.0 and $V = 0.05$, and two additional terms A and B dealing with the uncertainties and approximations involved in idealizing the dead and live load for the analysis. The latter account for such things as considering live loads to be uniformly distributed, static loads, idealizing localized loads as point loads, etc., These are assumed to have a mean of 1.0 and $V_A = 0.04$ und $V_B = 0.2$ for factors affecting the idealization of the dead and live load, respectively.

The analyses in MacGregor's (1979) report was based on the assumption that the structural analysis effects will be larger for live loads than for dead loads. No attempt will be made to distinguish between the type of structure or member, however. This is based on the assumption that although the accuracy of an analysis of an indeterminate structure is probably less than that of a statically determinate beam, there is more potential for load redistribution in such a structure and this offsets much of the loss of safety due to possible inaccuracies in the analysis. The factor E will be

Loads and Structure Failure

assumed as follows: for dead load $\mu_{ED} = 1.0$ $V_{EO} = 0.08$ and for live load $\mu_{EL} = 1.0$ and $V_{EL} = 0.20$.

The load for the reliability-based design was developed by Ravindar et al. (1974) for the reinforced concrete beam as the dead and live load effects are considered in developing the ultimate bending capacity of the beam. The dead load is relatively constant in the service life of the structure. The live load is the maximum value that occurs in the life time of the structure.

The mean dead load action, μ_D however, in any code the specified value by using the unit weight given in the code but Rosenblueth (1974) reported some major deviations as these may be happened due to significant alterations in the structures and not envisaged at the time of design. The dead load variation in this research consists of two uncertainty as V_{Do} depends on geometric factors that are estimated to be 0.04. The second variation V_E is the dispersion in the load effect prediction and is equal to 0.1. Therefore, the dead load variation is equal to 0.108 based on the following equation:

$$V_D = \sqrt{V_{DO}^2 + V_E^2} \tag{5.19}$$

The maximum live load intensity along the structure life time depends on the following factors:

1. Type of occupancy
2. Tributary area
3. Projected service life

A satisfactory theoretical model taking into account all these factors to predict the parameters of live load is not yet available. Though an otherwise comprehensive study was discussed by Pier in 1971. The mean μ_L and coefficient of variation V_{LO} of the lifetime maximum office live loads as a functions of the tributary area and number of tenancies are given in Tables 5.9 and 5.10. As in the case of dead load, the dispersion in the live load effect prediction is taken as $V_E = 0.1$.

The coefficient of variation of the live load effect is then given by

$$V_L = \sqrt{V_{LO}^2 + V_E^2} \tag{5.20}$$

TABLE 5.9
Mean of Life Time Maximum Office Live Load, N/m²

Number of Tenancies	Tributary Area, m²			
	4.6	32.6	74.4	112
1	647	608	598	589
2	939	805	772	748
5	1320	1030	982	938
10	1600	1200	1150	1080

TABLE 5.10
CoV of Life Time Maximum Office Live Loads

	Tributary Area, m²			
Number of Tenancies	4.6	32.6	74.4	112
1	0.83	0.57	0.52	0.48
2	0.47	0.36	0.34	0.32
5	0.35	0.30	0.28	0.27
10	0.34	0.28	0.27	0.27

5.6.2 Live Load Causing Failure

In some countries, there are controls on building occupancy as if it is used only for residential buildings and have a license by that so it cannot be converted its use to be a hospital or even administration buildings for example. However, in some countries there is no control at all; as in developing countries, for the building occupancy therefore in this situation, the probability of failure is higher due to change on building use without any management of change policy.

5.7 WIND LOAD STATISTICS

Katsura et al. (1992) reported that 80% of the 63 fatalities due to Typhoon Mireille, which struck Japan in 1991, were male, and more than 70% were over 60 years old. The causes of death were 31% blown by wind, 30% caught under collapsed or blown obstacles, and 23% hit by wind-borne debris. Fukumasa (1992) reported interesting statistics on the number of fatalities in Japan due to 114 typhoons in the 40 years from 1951 to 1990. The number of fatalities significantly reduced in more recent years, and this might be attributed to improved wind resistant design and construction technologies and also to development and improvement of weather forecasting methods and information transferring technologies. However, wind-induced damage to buildings and structures, agricultural and forestry products, and so on are still very severe. The property insurance money paid was almost 6 billion US$ for Typhoon Mireille in September, 1991, 18 billion US$ for Hurricane Andrew in August, 1992, and almost 8 billion US$ for 10 typhoons that attacked Japan in 2004. Hurricane Katrina killed 2,541 people in August, 2005 and caused 28 billion US$ economic loss in the US, Cyclone Sidr in November, 2007 killed 4,234 people and caused 1.7 billion US$ economic loss in Bangladesh, and Cyclone Nargis in May, 2008 killed 138,366 people and caused 10 billion US$ economic loss. Very recently, Typhoon Morakot struck Taiwan and neighboring countries in August, 2009, and fatalities and missing numbered 732. It recorded 2,888 mm integrated rainfall over 4 days in Chiayi Province, Taiwan. Almost all of these disasters resulted from the combined effects of strong wind and accompanying water hazards due to heavy rain and storm surge. The social impacts caused by these "wind-related" disasters were some of the most severe in our human society (Figure 5.8).

Loads and Structure Failure

FIGURE 5.8 Collapse of Power Transmission Towers (Typhoon Higos, 2002).

The wind pressure on the building in Section 6 of ASCE 7-95 for ordinary buildings is determined from the following equation:

$$W = q \cdot G \cdot C \cdot A \qquad (5.21)$$

where the velocity pressure, q, is determined as

$$q = 0.00256 K_z V^2 \qquad (5.22)$$

where V is equal to 50-year mean recurrence interval wind speed (3 s gust), K_z is the exposure factor, G is the gust factor, C_p is pressure coefficient—the topographic factor K_{zt} is neglected for simplicity and importance factor I equals 1.0 for ordinary buildings—and A is the projected area normal to wind.

The importance factor adjusts V for other mean recurrence intervals in accordance with perceived building risk. Maps and tables in the Standard provide specific values of parameters in Equation 5.21 for design.

The parameters V, K_z, G, and C_p (or GC_p) are random variables, and the probability distribution function of wind pressure and the wind load statistics are required to determine appropriate probability-based load factors and load combinations. The CDF of wind speed is particularly significant because V is squared in Equation 5.22. However, the uncertainties in the other variables also contribute to the uncertainty in W, and any analysis of wind load uncertainty that ignores these other sources will yield an unconservative and erroneous estimate of reliability under wind load.

The CDFs for the random variables in Equation 5.12 used to derive the wind load criteria that appear in ASCE 7-95 are summarized in Table 5.11 (Ellingwood, 1981). Because the reliability analyses were performed for a 50-year service period, the wind speed, V, in Table 5.11 is the 50-year maximum fastest mile wind speed, which can be determined from the CDF of the annual extreme fastest mile wind speed as discussed by Simiu and Filliben (1976) and Simiu et al. (1979) using order statistics. This wind speed is site dependent; seven stations were used to determine typical statistics, all of which were situated in non-hurricane prone regions.

TABLE 5.11
Wind Load Statistics for ASCE 7-95 Load Criteria (Ellingwood, 1981)

Parameter	Mean/Nominal	CoV	CDF
Exposure factor (Kz)	1.0	0.16	Normal
Gust factor (G)	1.0	0.11	Normal
Pressure coefficient (Cp)	1.0	0.12	Normal
50 years maximum fastest mile wind speed (V)	1.0	0.12	Type I
Wind pressure (W)	0.78	0.37	Type I

On the other hand, the CDF for pressure coefficient Cp was determined from both wind tunnel and field instrumentation as studied by Peterka and Cermak (1976), Marshall (1977), Cook and Mayne (1979), and Stathopoulos (1980). Information on G and on Kz (Davenport et al., 1979; Ravindra et al., 1978; Hart and Ellingwood, 1982) relied, in part, on judgment. The effect of wind directionality was reflected by a multiplicative factor 0.85, as noted previously.

The CDF of W was determined by numerical integration for each station and was fitted by a Type-I distribution of largest values in the 90th percentile and above—the region that is significant for structural reliability analysis.

It is found that the wind pressure can be presented by probability distribution extreme Type I with bias factor (mean/nominal) equal to 0.78 with coefficient of variation equal to 0.37.

Additional research during the past 18 years has provided additional information that can be used to confirm or revise these wind load statistics, as appropriate.

A search of the wind engineering literature to collect and synthesize this information for code purposes might omit important sources of information (e.g., wind tunnel studies not published in the archival literature) and introduce prejudices in the synthesis of the data. An alternative approach is to impanel a group of individuals familiar with wind engineering issues to supplement the above references with more recent studies that would (or would not) support them as sources of data and participate in a Delphi to revise the uncertainty measures that are the basis for the CDF of W and for the load factors. This approach has the advantage of drawing efficiently on the collective wisdom of individuals knowledgeable in the field.

5.8 EARTHQUAKE LOAD

There are many buildings which collapsed worldwide due to seismic load. As seismic load occurs suddenly its magnitude is not known (Figures 5.9–5.11).

The peaks of a Gaussian process have been shown to be modeled well by the Rayleigh distribution.

Furthermore, the response of a non-Gaussian process, such as the peaks of structural response during earthquakes, has been shown to be modeled well by a Weibull distribution. This is somewhat logical since the Rayleigh distribution is a special case of the Weibull distribution, and a change in the shape parameter provides significant modeling flexibility. The two-parameter Weibull distribution was used in this

Loads and Structure Failure 111

FIGURE 5.9 Large shear movement due to seismic load in Pakistan.

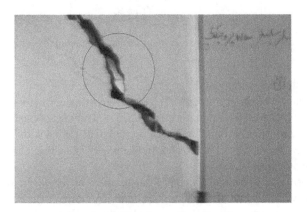

FIGURE 5.10 Cracks due to differential settlement.

study to model the statistical distribution of the deformation response peaks and has cumulative distribution function:

$$F(x) = 1 - \exp\left\{-\left(\frac{x}{\lambda}\right)^k\right\} \tag{5.23}$$

where x is the random variable of interest and k and j are the scale and shape parameters, respectively. In the present study, k and j were treated as random variables in order to describe an entire family of earthquake response peaks. This family of response peaks was felt to be representative of the suite of parent earthquakes (i.e., suite of 10 earthquakes), discussed earlier, from which they were derived. In order to do this, the Weibull distribution as shown in Equation 5.23 can be re-written as

$$F(x) = 1 - \exp\left\{-\left(\frac{x}{\lambda\theta}\right)^{k\varphi}\right\} \tag{5.24}$$

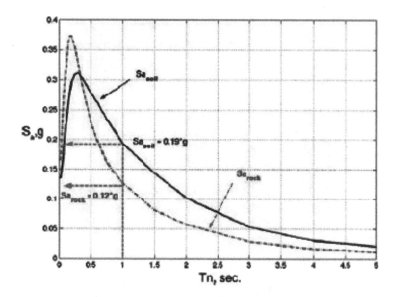

FIGURE 5.11 Response spectra for an earthquake having magnitude 6.5.

where Θ and Φ are random variables each having a mean of unity, and a standard deviation calculated from the dataset. The elasto–plastic deformation was idealized the structure as studied by Lindt and Goh (2004). The elasto–plastic deformation peaks, both Θ and Φ, were found to fit a lognormal distribution with 95% confidence using a K–S test as stated by Ayyub and McCuen (1997); hence, a lognormal model was felt to be adequate.

The lognormal distribution is defined on the interval $[0,\infty]$. For very small values of Φ, the value of the cumulative distribution function, $F(x)$, in Equation 5.24 becomes very small and may cause numerical problems. It can also be reasoned that values outside of the original elasto–plastic responses would contradict the seismic hazard analysis logic used in development of the ground motion suites. Hence, the lower and upper tail of the lognormal distribution for h and u were truncated so that the minimum and maximum h and u values were equal to the minimum and maximum from the deformation response peak data set. Mathematically, the new distribution, f(y), can be expressed as stated by Cornell (1970):

$$F(y) \begin{array}{ll} = 0 & \text{for } y < \min(y) \\ = kf(x) & \min(y) \leq y < \max(y) \\ = 0 & y \geq \max(y) \end{array} \quad (5.25)$$

where

$$k = \frac{1}{F(\max(y)) - F(\min(y))} \quad (5.26)$$

Finally, the earthquake demand in terms of the peaks of the elasto–plastic response can be expressed by Equation 5.24 for each combination of elasto–plastic oscillator and suite of earthquakes.

For loads on bridges, in the case of applying a reliability analysis for the bridges it need some care as the major loads are coming from the moving loads of cars and trucks. In addition, the dead load has a higher value than that in the normal building. As the guideline is a study which is performed by Stewart et al. (2001) as in his case the mean value of the paving thickness is 90 mm with coefficient of variation 0.25 and for the dead load with bias factor 1.05 and coefficient of variation 0.1 and presented by a normal distribution for the live load due to single, heavy-loaded trucks with mean load of 275 kN, the CoV is 0.41, as presented by the normal distribution.

BIBLIOGRAPHY

Allen, D.E., 1975, Limit states design—a probabilistic study, *Canadian Journal of Civil Engineering*, Vol. 2, pp. 36–49.
ANSI 58.1-1972, Building code requirement for minimum design loads in buildings and other structures, American National Standard Institute, New York, 1972.
ANSI 58.1-1980, Building code requirement for minimum design loads in buildings and other structures, American National Standard Institute, New York, 1980.
Ayyub, B.M., and McCuen, R.H., *Probability, Statistics, and Reliability for Engineers*, CRC Press LLC, Boca Raton, FL, 1997.
Benjamin, J.R., and Cornell, C.A., *Probability, Statistics, and Decision for Civil Engineers*, McGraw Hill, New York, 1970.
Bolotin, V.V., *Statistical Methods in Structural Mechanics*, Holden day, San Francisco, CA, 1969.
Bosshard, W., On stochastic load combination, Technical Report No. 20, J.A. Blume, Earthquake Engineering Center, Stanford, CA, 1975.
Bryson, J.O., and Gross, D., Techniques for the survey and evaluation of live floor loads and fire loads in modern office buildings, U. S. Department of Commerce, National Bureau of Standards, Building Science series, Washington, D.C., Dec. 1968.
Cartwright, D.E., and Longuet-Higgins, M.S., 1956, The statistical distribution of maxima of a random function, *Proceedings of the Royal Society of London*, Vol. A237, pp. 212–32.
Choi, E.C.C., 1990, Live load for office buildings: Effect of occupancy and code comparison, *Journal of Structural Engineering*, Vol. 116, No. 11, pp. 3162–74.
Cook, N.J., and Mayne, J.R., 1979, A novel working approach to the assessment of wind loads for equivalent static design, *Journal of Wind Engineering and Industrial Aerodynamics*, Vol. 4, pp. 149–64.
Corotis, R.B., 1972, Statistical analysis of live load in column design, *Journal of the Structural Division*, Vol. 98, No. ST8, pp. 1803–15.
Corotis, R.B., and Chalk, P.L., 1980, Probability models for design live loads, *Journal of the Structural Division*, Vol. 106, No. ST10, pp. 2017–33.
Corotis, R.B., and Doshi, V.A., Stochastic analysis of floor loads, *Proceedings of the National Structural Engineering Specialty conference, ASCE*, Madison, Aug. 1976, pp. 72–94.
Corotis, R.B., and Doshi, V.A., 1980, Probability models for live load survey results, *Journal of the Structural Division*, Vol. 106, No. ST10, pp. 2017–33.
Corotis, R.B., and Tsy, W-Y., 1983, Probabilistic load duration model for live loads, *Journal of the Structural Division*, Vol. 109, No. 4, pp. 859–873.

Culver, C., Survey results of fire loads and live loads in buildings, Building Science Series 85, U.S. Department of Commerce, National Bureau of Standards, Washington, D.C., May 1976.

Culver, C.G., 1976, Live load survey results for office buildings, *Journal of the Structural Division*, Vol. 102, No. ST12, pp. 2269–2284.

Culver, C., and Kusher, J., A program for survey of fire loads and live loads in building, Technical Note 858, U.S. Department of Commerce, National Bureau of Standards, Washington, D.C., May 1975.

Davenport, A.G., Isyumov, N., and Surry, D., The role of wind tunnel studies in design against wind, *Proceedings of the ASCE Spring Convention*, ASCE, Reston, VA, 1979.

Dunham, J.W., Brrekke, C.N., and Thompson, G.N., Live loads on floors buildings, Buildings and Structures Report 133, U.S. National Bureau of Standard Washington, D.C., Dec. 1952.

Euro-International Committee for Concrete, Comité Euro-International du Béton, *Basic Data on Loads, CEB International Course on Structural Concrete, Part Cl-2*, Laboratorio Nacional de Engenharia Civil, Lisbon, 1973.

Fukumasa, Y., 1992, Change in the number of casualties due to typhoon striking, *Study on Disasters*, Vol. 23, 84–89 (in Japanese).

Galambos, T.V., and Ravindra, M.K., Tentative load and resistance factor design criteria for steel buildings, Research Report No. 18, Civil and Environmental Engineering Department, Washington University, St. Louis, MO, 1973.

Hart, G., and Ellingwood, B., 1982, Reliability-based design considerations for relating loads measured in wind tunnel to structural resistance, In: *Wind Tunnel Modeling for Civil Engineering Applications*, T.A. Reinhold, (ed.), Cambridge University Press, New York, pp. 27–42.

Hasofer, A.M., 1968, Statistical model for live floor loads, *Journal of the Structural Division*, Vol. 94, No. ST10, pp. 2183–96.

Katsura, J., Taniike, Y., and Maruyama, T., 1992, Damage due to Typhoon 9119 (human damage), Study on strong-wind disasters due to Typhoon No. 19 in 1991, pp. 91–94 (in Japanese).

Lind. N.C., Approximate analysis, safety and economics of structures. (In Press. ASCE).

Marshall, R.D., 1977, The measurement of wind loads on a full-scale mobile home, *Rep. NBSIR 77-1289*, National Bureau of Standards, Washington, D.C.

McCabe, P.M., et al., Data processing and data analysis procedures for fire loads and live loads survey program, NBSIR 76-982, U.S. Department of Commerce, Washington, D.C., 1975.

Mitchell, G.R., and Woodgate, R.W., A survey of floor loading in office buildings, London, England, Aug. 1971.

Mitchell, G.R., and Woodgate, R.W., Floor loadings in offices - the results of a survey, Current Paper 3/71, Department of Environment Building Research Station, Garston, England, 1971.

Mitroff, I.I., and Tuoff, M., 1975, Philosophical and methodological foundation of Delphi, In: *The Delphi Method, Techniques and Applications*, H.A. Linstone, and M. Turoff, (eds.), Addison-Wesley, Reading, MA, pp. 17–36.

Murphy, J.F., Ellingwood, B.R., and Henrickson, E.M., 1987, Damage accumulation in wood structural members under stochastic live load, *Wood and Fiber Science*, Vol. 19, No. 4, pp. 453–63.

National Building Code, 1975, National Research Council, Ottawa, Canada.

Niedzwecki, J.M., van de Lindt, J.W., Gage, J.H., and Teigen, P.S., 2000, Design estimates of surface wave interaction with compliant deepwater platforms, *Ocean Engineering*, Vol. 27, 867–88.

Peterka, J., and Cermak, J.E., 1976, Wind pressures on buildings probability densities, *Journal of the Structural Division*, Vol. 101, No. 6, pp. 1255–1267.

Ravindra, M.K., Lind, N.C., and Slu, W., 1974, Illustrations of reliability based design, *Journal of the Structural Division*, Vol. 1, No. ST9, pp. 1789–811.

Rosowky, D.V., and Ellingwood, B.R., Stochastic damage accumulation and probabilistic codified design for wood, Civil, Engineering, Rep No. 1990-02-02, Johns Hopkins University, Baltimore, MD, 1990.

Somerville, P., Smith, N., Punyamurthula, S., and Sun, J., Development of ground motion time histories for phase 2 of the FEMA/SAC steel project, SAC Background Document Series Final Report, Sacramento, CA, 1997.

Stewart, M.G., Rosowsky, D.V., and Val, D.V., 2001, Reliability-based bridge assessment using risk-ranking decision analysis, *Structural Safety*, Vol. 23, pp. 397–405.

Wen, Y.K., Probability of extreme loads, *Proceeding Fourth International Conference on Structural Mechanics in Realtor Technology*, San Francisco, CA, Aug. 1977.

Wen, Y.K., 1977, Statistical combination of extreme loads, *Journal of the Structural Division*, Vol. 103, No. ST5, pp. 1079–93.

6 Reliability-Based Design for Structures

6.1 INTRODUCTION

The structural reliability is the reverse meaning for structural probability of failure. This chapter presents some of the theoretical backgrounds relating to safety. The civil engineers in the university are using the equation of design with the factors. This is called deterministic study. But from research point of view the researchers are using a probabilistic study to define the factors of safety in the codes.

For example, the factor of dead load in BS is 1.4, while in European code it is 1.35. There are a lot of researches and development works to reduce this number as these safety factors affect the country economic and do not look only to your project. The reliability analysis for reinforced concrete structure will be discussed in this chapter. Note that there are many researches covering different concrete structure element reliability.

Frangopol et al. (2000) discussed the performance of the slender reinforced concrete columns under random loads as in designing the column usually considered that the axial load and pending moments are perfectly correlated. In real life the axial load and bending moment is a result of dead load and live load; however, in some cases the maximum bending moment is a result of the wind load as the wind load effect is not correlated with the dead load or live load.

This study shows that the reliability of RC columns depends on the loading history of the column. Correlation between the loads affects the column's reliability, too. The effects of correlation vary depending on the slenderness of the column and the region of the interaction diagram in which the column is loaded. For high slenderness, perfect positive correlation between axial load and bending moment proves to be the most conservative condition, regardless of the region of the interaction diagram in which the column in loaded. For columns with low-to-moderate slenderness ratios subject to high axial load, perfect positive correlation between loads is the most conservative assumption. When subjected to low axial loads, i.e., below the balanced point, this assumption proves to be unconservative. These are effects that cannot be captured in conventional code-based analysis as ACI 318.

6.2 RELIABILITY OF REINFORCED CONCRETE COLUMN

In this chapter, the reliability analysis of the reinforced concrete column is discussed. Parametric study of the reliability of column in residential building is performed.

The straining actions at the column base are calculated by doing space structural analysis using SAP90 program due to the live load model described in Chapter 3.

The strength of the reinforced concrete column is calculated according to the Egyptian code and approximate equations derived by Ahmed (1985).

The limit state equation is formulated and the probability of failure is calculated using Monte-Carlo simulation technique.

The parameters that are studied are as follows: the location of the column, eccentricity, concrete strength, dead load, and steel strength. To calculate the reliability of the reinforced concrete column, the straining action should be calculated in every member and in here we are focusing on the column as it is the most critical member on the building as it is governing the collapse of the building. The ultimate strength of the reinforced concrete column is calculated and, based on that, one can formulate the limit state equation of the reinforced concrete column. There are different parameters that affect the probability of failure of the reinforced concrete column are applied on a building. Therefore, the probability of failure will be calculated in different cases to present the results and observations on studying different parameters effects on the reliability of the columns.

6.3 CALCULATION OF THE STRAINING ACTIONS AT THE COLUMN BASE

A three-dimensional structure analysis will be performed for the structure considering only a unit load (1 t/m²) acting on the jth bay of the ith floor.

The calculated straining actions at the column base in this case will be as follows:

N_{ij} is the normal force at the column base due to unit distributed load acting on the jth bay of the ith floor,

M_{xij} is the bending moment about x-axis at the column base due to unit distributed load acting on the jth bay of the ith floor, and

M_{yij} is the bending moment about y-axis at the column base due to unit distributed load acting on the jth bay of the ith floor.

Considering the actual value of the load W_{ij} acting on the jth bay of the ith floor, the straining action at the column base can be obtained by multiplying N_{ij}, M_{xij}, and M_{yij} by the load value W_{ij}.

The total values of the straining actions at the column base due to loading different bays and floors can be determined as follows:

$$N = \sum_{i=1}^{n_1} \sum_{j=1}^{n_2} N_{ij} W_{ij} \tag{6.1}$$

$$M_x = \sum_{i=1}^{n_1} \sum_{j=1}^{n_2} M_{xij} W_{ij} \tag{6.2}$$

$$M_Y = \sum_{i=1}^{n_1} \sum_{j=1}^{n_2} M_{xij} W_{ij} \tag{6.3}$$

Reliability-Based Design for Structures

TABLE 6.1
Statistical Parameters and Distributions Assumed in the Parametric Study

Design Variables	Specified	Mean in Situ	Mean in Situ/Specified	σ	CoV	Distribution
F_{cu} kg/cm²	200	197	–	35.46	0.18	Normal
F_y kg/cm²	2,400	2,640	1.1	184.8	0.07	Normal
t (cm)	50	50.2	1.004	1.255	0.025	Normal
B (cm)	50	50.2	1.004	1.255	0.025	Normal
A_s (cm²)	24.127	23.64	0.98	0.71	0.03	Normal
L (kg/m²)	200[a]	106.13	–	42.52	0.4	Lognormal
D (kg/m²)	–	–	1.05	–	0.1	Normal
E	–	–	1	–	0.05	Normal
A	–	–	1	–	0.04	Normal
B	–	–	1	–	0.2	Normal
R_f	–	–	0.98	–	0.05	Normal

[a] According to EC-89.

where n_1 is the total number of floors and n_2 is the total number of bays per floor. W_{ij} is the actual live load effect at floor number i and bay number j. The values of N, M_x, and M_y at each bay will be calculated by modeling the structure by structure analysis.

The live load is taken as described in Chapter 3. According to Gallambos (1982), the dead load is presented by a normal distribution with bias factor (ratio of mean to nominal) equal to 1.05 and coefficient of variation (CoV) equal to 0.1.

The load effect, S, for combined dead and live loads will be assumed to have the form (Galambos 1978):

$$S = E(A \cdot D_L + B \cdot L_L) \quad (6.4)$$

where D_L and L_L are random variables representing the straining action of the dead and live loads, respectively. A and B are random variables reflecting the uncertainties in the transformation of load into load effects and E is a random variable representing the uncertainties in structural analysis.

In order to calculate the values of load, L, on the columns, the statistical values of the variables in Equation 6.4 are considered as shown in Table 6.1.

6.4 ULTIMATE STRENGTH OF REINFORCED CONCRETE COLUMNS

The columns considered in this thesis are short, tied, symmetrically reinforced, and have rectangular cross-sections.

The ultimate strength of a column is the sum of the forces required to bring concrete to crushing and steel to yielding.

The ultimate strength is calculated based on the assumption that the entire concrete cross-section is uniformly stressed to 0.67 the concrete compressive strength and the strain is 0.002 in axial compression and 0.003 in cases of small eccentricity.

6.4.1 UNIAXIALLY LOADED COLUMN

In general, a column subjected to an axial eccentric force causes a uniaxial moment, as shown in Figure 6.1, and fails when the concrete reaches crushing.

In the analysis of buildings due to vertical loads only, columns are usually subjected to axial force with very small eccentricity where the entire column cross-section is subjected to compressive stress. Therefore, the case of very small eccentricity will be discussed in this section.

In this case, the strain distribution and the corresponding stresses in steel and equivalent stresses in concrete are shown in Figure 6.1. For very small eccentricity, the corresponding stress resultants carried by the cross-section element, as shown in Figure 6.1, are as follows:

$$C_c = 0.67 F_{cu} \cdot b \cdot t$$
$$C_s = A_s \cdot F_s \qquad (6.5)$$
$$C'_s = A_s \cdot F_y$$

where

C_c is the resultant force in concrete,
C_s is the resultant force in steel in tension side of moment, and
C'_s is resultant force in steel in other side

In case of very small eccentricity, all of the sectional area is assumed to be in compression and the steel in the tension side is near the yield strength.

From equilibrium of internal and external moments about the tension steel then.

$$R_u(e + d_1/2) = C_c(d_1/2) + C'_s \cdot d_1 \qquad (6.6)$$

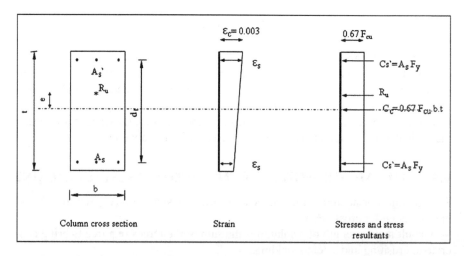

FIGURE 6.1 Strains, stresses, and stress resultants at ultimate in a section subjected to eccentric compression with very small eccentricity.

Reliability-Based Design for Structures

Substituting C_c, C_s, and C'_s from Equation 6.5 with Equation 6.6 for solving for R_u, one can obtain

$$R_u = \frac{0.67 F_{cu} \cdot b \cdot t(d_1/2) + (A_s F_y) \cdot d_1}{(e + d_1/2)} \quad (6.7)$$

where

b is the width of the column,
t is the length of the column,
F_y is the steel bar yield strength,
F_{cu} is the concrete compressive strength,
A_s and A'_s are the steel bar area for tension side and other, and
e is the eccentricity of load on the column which is calculated by

$$e = M/N$$

where M and N are the bending moment and normal force calculating from the straining action. But, in this study, there are biaxial moments acting on the column, so the case of biaxial moment will be discussed.

The resistance of the reinforced concrete column, R, is calculated as follows:

$$R = R_u \cdot R_f \quad (6.8)$$

where R_u is the average value of strength calculated from Equation 6.7 and R_f is a value represent the ratio between the actual capacity of the member and the calculated member capacity as this factor is due to the accuracy of the code equation.

The statistical parameters of this value are presented in Table 6.1.

6.4.2 BIAXIALLY LOADED COLUMN

Several investigators have suggested empirical approximations or design charts to represent the failure surface in case of biaxial bending.

There are two analytical expressions for the interaction surface suggested by Bresler (1960). The first, which is widely known as Bresler's reciprocal equation, is

$$\frac{1}{P} = \frac{1}{P_X} + \frac{1}{P_Y} - \frac{1}{P_o} \quad (6.9)$$

where P is the required ultimate load in the presence of two given bending moment, P_x is the ultimate load in the presence of the moment about the x-axis, P_y is the ultimate load in the presence of the moment about the y-axis, and P_o is the concentric ultimate load.

The second equation is given as follows:

$$\left(\frac{M_X}{M_{XO}}\right)^\alpha + \left(\frac{M_Y}{M_{YO}}\right)^\beta = 1 \quad (6.10)$$

where M_x and M_y are the failure moments if applied together in the presence of a certain axial force, M_{xo} and M_{yo} are the failure moments if applied separately in the presence of the same axial force, and α, β are coefficients which depend on the geometry of the section and the properties of concrete and steel. Meek (1963) suggested a bilinear relation between M_x and M_y and verified that relation with test results. Furlong (1979) showed that Bresler's relation between moments and moment capacities, with each of α, β taken equal to 2, was in good agreement with test results. Fintel (1974) suggested that each of α and β may be taken equal to unity.

It is worth to mention that the numerical solution of the problem of designing a reinforced concrete section subjected to biaxial bending is very tedious and complicated since it involves successive iterations based on assumptions of the location of the neutral axis.

Therefore, Egyptian code EC-89 presents a simplified design method for the common case of a symmetrically reinforced rectangular section.

The method simplified permits the design of the section to be carried out as if it were subjected to an increased moment about one axis given by the following equations (Abd El-Rahman 1993; Figure 6.2):

$$M'_x = M_x + B_c(t'/b')M_y \quad \text{if} \quad M_x/t' \geq M_y/b \tag{6.11}$$

$$M'_y = M_y + B_c(b'/t')M_x \quad \text{if} \quad M_x/t' < M_y/b' \tag{6.12}$$

where M'_x and M'_y are the effective uniaxial design moment about x- and y-axis, respectively.

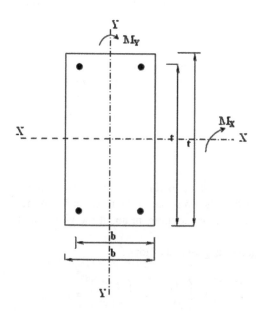

FIGURE 6.2 Biaxial bending.

The coefficient B_c is calculated as follows:

$$B_c = 0.3 + 1.167\left(0.6 - \left(P_u/t.b.F_{cu}\right)\right) \geq 0.30 \qquad (6.13)$$

6.5 LIMIT STATE EQUATION AND RELIABILITY ANALYSIS

The limit state equation is a function of the resistance, R, of the structural element and of the load effect, L, acting on it; R and L are random variables:

$$Z = R/L \qquad (6.14)$$

Since R and L are random variables, the Z is also random variables.

R and L are calculated from Equation 6.8 and Equation 6.4, respectively.

Knowing the probability density function of the resistance, R, and load effect, L, the probability density function of Z can be obtained.

The probability of failure of a structural element (the shaded area), represented in Figure 6.3, is equal to

$$P_f = P[R/L < 1.0] \qquad (6.15)$$

Since ln1.0=0.0, then Equation 6.15 becomes

$$P_f = P\left[\ln(R/L) < 1.0\right],$$

$$P_f = \int_0^1 f_z(z)\,dz = F_z(1.0) \qquad (6.16)$$

where $f_z(z)$ is the probability density function of the random variable Z.

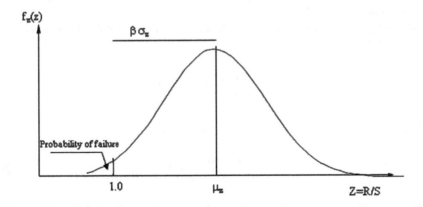

FIGURE 6.3 Definition of safety index and probability of failure.

If the probability density function of variable, z, is a normal distribution, the probability of failure will be calculated by the following equation:

$$P_f = P[Z < 1.0] = \phi\left[\frac{1.0 - \mu_z}{\sigma_z}\right] \quad (6.17)$$

where ϕ is the cumulative standard normal distribution function.

Safety can be measured in terms of a "safety index", β, which was defined by Cornell (1969) using the second-moment format as the number of standard deviations of the safety factor z by which its mean exceeds 1.0 as shown in Figure 6.3.

This, the safety index, β, is

$$\beta = \frac{1 - \mu_z^N}{\sigma_z^N} \quad (6.18)$$

According to Ang and Tang (1984), Z is normally distributed random variable if R and S are lognormally distributed variables.

In this study, the live load is modeled by a lognormal distribution. However, the strength of structural member has different parameters for different probability distributions. So, the random variable, R, representing the strength of the structural element may be modeled by a distribution not lognormal distribution. Therefore, the random variable, Z, representing the safety of the structural element will not be a normal distribution.

Hence, the random variable, Z, will be converted to an equivalent normal distribution which will be described in the following section.

6.5.1 Equivalent Normal Distribution

If the probability distribution of the random variable Z is known, the probability P_f may be evaluated by numerical integration technique. The equivalent normal distribution for a non-normal variant may be obtained such that the cumulative probability as well as the probability density ordinate of the equivalent normal distribution are equal to those of the corresponding non-normal distribution at the appropriate point on the failure surface (Ang et al. 1984).

Equating the cumulative probabilities as described above at the failure point with Z = 1.0, one can obtain

$$\phi\left(\frac{1.0 - \mu_Z^N}{\sigma_Z^N}\right) = F_Z(1.0) \quad (6.19)$$

Substituting Equation 6.16 into Equation 6.19 yields

$$\phi\left(\frac{1.0 - \mu_Z^N}{\sigma_Z^N}\right) = P_f \quad (6.20)$$

where μ_Z^N and σ_Z^N are the mean value and standard deviation, respectively, of the equivalent normal distribution for Z.

Reliability-Based Design for Structures

The above equality then yields

$$\mu_Z^N = 1.0 - \sigma_Z^N \phi^{-1}[P_f] \tag{6.21}$$

Equating the corresponding probability density ordinates at $Z = 1.0$, one can obtain

$$\frac{1}{\sigma_Z^N} \phi\left(\frac{1.0 - \mu_Z^N}{\sigma_Z^N}\right) = f_z(1.0) \tag{6.22}$$

Solving Equations 6.20 and 6.22, the standard deviation of the equivalent normal distribution, Z, can be obtained as

$$\sigma_Z^N = \frac{\phi\{\phi^{-1}[P_f]\}}{f_z(1.0)} \tag{6.23}$$

Thus, using Equations 6.21 and 6.23, one can obtain μ_Z^N, σ_Z^N.
Then the reliability index is

$$\beta = \frac{1 - \mu_Z^N}{\sigma_Z^N} \tag{6.24}$$

On the other hand, Equation 6.24 can be used to evaluate the probability of failure P_f by converting the non-normal distribution variant into normal distribution variant.

6.6 PARAMETERS AND METHODOLOGY

The three-dimensional structure analysis will be performed to determine the straining actions at the column base. The live load will be considered as a lognormal distribution with statistics parameters as described in Chapter 4.

In the analysis to determine the straining actions at the column base, the live load is assumed on only one bay of one floor with a value 1 t/m². The straining actions, N_{ij}, M_{xij}, and M_{yij}, are calculated at the base of the columns at the considered locations. The above analysis is repeated considering live load on different bays and floors. The obtained values of the straining actions at the base of the different columns due to loads acting on different positions (load effect) were calculated by Elreedy et al. (2000).

The reliability is calculated using an Excel spreadsheet, and the required Monte-Carlo simulation is performed using Crystal ball program (1996).

A computer program is prepared using Excel to calculate the reliability of the reinforced concrete short column using Monte-Carlo simulation technique. This program is based on the limit state equation given as discussed before.

The calculation procedure of reliability index of columns will be obtained by the following steps:

1. Generate random values for all the design variables using the Crystal ball program. This generation includes random values for live load acting on each bay/floor of the building.

2. Calculate the straining actions at the base of the column N, M_x, and M_y due to live loads acting on the different bays and floors by summing the multiplication of the generated value of live load in each bay/floor by the corresponding coefficient of load value effect due to 1 t/m^2.
3. Calculate the total straining actions at the column base due to dead load and live load combined by summing the straining actions due to dead load and those due to live load calculated in step 2.
4. Calculate the column capacity from Equation 6.8.
5. Calculate the value of the safety factor, Z, by substituting the generated random values of the design variables into the limit state equation.
6. Perform 10,000 trials for the steps 1–4.
7. Based on the 10,000 value of Z obtained in step 6, the Crystal ball program plots a histogram for z and approximates this histogram to a suitable distribution of the famous probability distribution.
8. Knowing the statistical parameters and distribution of Z from step 7, calculate the equivalent normal parameters for Z.
9. Calculate the failure probability and safety index of the column using Equations 6.17 and 6.24, respectively.

The previous procedure is performed for the different column locations. Moreover, for each case, different ratios between length and width (t/b) are used to calculate the column reliability.

6.7 APPLICATION ON A BUILDING

In this study, a six-floor building is considered, with each floor consisting of four bays. Two bays are in the x-direction and two bays are in the y-direction.

The values of live load that apply on each bay are assumed to be independent.

The floor height of the building, H, is 3.0 m. The bay span in the x-direction, Lx, is 5.0 m and the bay span in the y-direction, Ly, is 4.0 m. The statistical parameters that are used in this study to calculate the reliability of the reinforced concrete column are summarized in Table 6.1.

Different column locations and dimensions are considered. These locations include, as shown in Figure 6.1, interior column (column I) and edge columns (columns E1, E2). The effect of these locations on the reliability of column is discussed.

The three-dimensional structure analysis is performed by using a computer program package SAP90. The straining actions at the column base are calculated for different column locations and dimensions.

The straining actions at the column base of the different columns due to dead load effect are calculated considering the actual weight of the building elements. It is noticed that the straining actions due to dead load are mainly axial forces on the columns.

The reliability is calculated using an Excel spreadsheet, and the required Monte-Carlo simulation is performed using Crystal ball program (1996).

The reliability of the reinforced concrete columns is performed for the different column locations. Moreover, for each case, different ratios between length and width (t/b) are used to calculate the column reliability.

Reliability-Based Design for Structures

The reliability of reinforced concrete columns is also calculated assuming different values of eccentricity.

Moreover, the effect of major limit state variables as concrete strength, dead load, and steel strength is discussed.

6.8 EFFECT OF COLUMN LOCATION

The reliability index is determined for the columns of the considered building taking into account the different locations of these columns. Different cross-sections of different aspect ratio for each column are considered as well.

The results including the reliability index and the corresponding probability of failure for the different column cross-sections are presented in Tables 6.2–6.5 for intermediate (I), exterior (E1), other exterior (E2), and corner column (C), respectively.

From these tables, it is noticed that there is a slight difference in the reliability index of columns for different aspect ratios regardless of column location. However, interior columns of square section have smaller reliability index than those of rectangular sections.

To study the effect of column locations on reliability, square columns are only considered for all columns to avoid the effect of moment of inertia in the comparison.

The reliability of the columns of different locations is shown in Figure 6.4.

Figure 6.5 shows that the corner column has less reliability index. So, the corner column is the critical one as it has a higher probability of failure.

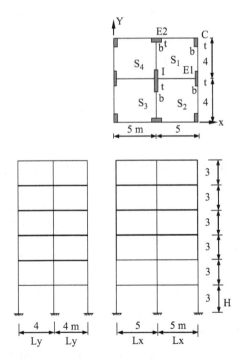

FIGURE 6.4 Structural model of the building considered.

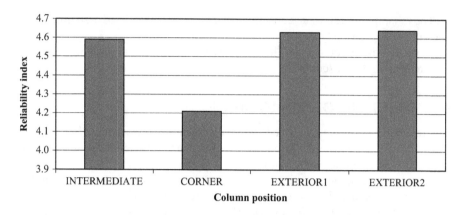

FIGURE 6.5 Relation between column position and reliability index.

TABLE 6.2
Relation between Reliability and Interior Column (I) Dimensions and Orientation

Section Dimension and Orientation	Length to Width Ratio t/b	Probability of Failure P_f	Reliability Index β
90 × 25	3.60	1.417E–06	4.68
65 × 25	1.86	1.668E–06	4.65
50 × 45	1.11	1.642E–06	4.66
50 × 50	1.00	2.171E–06	4.59
45 × 50	0.90	2.170E–06	4.59
35 × 65	0.54	1.888E–06	4.63
25 × 90	0.28	1.443E–06	4.68

TABLE 6.3
Relation between Reliability and Exterior Column (E1) Dimensions and Orientation

Section Dimension and Orientation	Length to Width Ratio t/b	Probability of Failure P_f	Reliability Index β
50 × 25	2.0	8.36E–07	4.77
35 × 35	1.0	1.84E–06	4.63
25 × 50	0.5	1.74E–06	4.65

6.8.1 Effect of Eccentricity

Reliability of eccentric columns is determined by considering the eccentricity in the range specified by the Egyptian code of practice.

TABLE 6.4
Relation between Reliability and Exterior Column (E2) Dimensions and Orientation

Section Dimension and Orientation	Length to Width Ratio t/b	Probability of Failure P_f	Reliability Index β
50 × 25	2.0	2.13E–06	4.60
35 × 35	1.0	1.76E–06	4.64
25 × 50	0.5	1.83E–06	4.63

TABLE 6.5
Relation between Reliability and Corner Column (C) Dimensions and Orientation

Section Dimension and Orientation	Length to Width Ratio t/b	Probability of Failure P_f	Reliability Index β
30 × 25	1.20	1.68E–05	4.14
30 × 30	1.00	1.27E–05	4.21
25 × 30	0.83	8.97E–06	4.28

In this case, the intermediate column is only considered. The reliability index versus the eccentricity is plotted as shown in Figure 6.6. From this figure, one can conclude that the reliability index of column is decreased gradually with increasing the eccentricity.

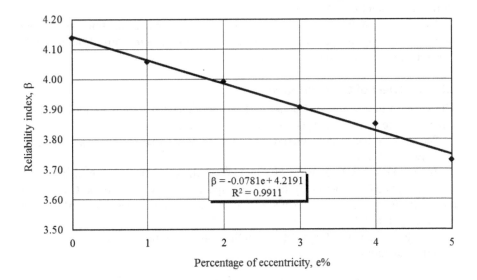

FIGURE 6.6 Relation between percentage of eccentricity and reliability index.

By using the regression analysis method, the relation between the reliability index and percentage of eccentricity (e %) can be approximated by the following equation:

$$\beta = 4.2191 - 0.0781e \qquad (6.25)$$

An approximate value of safety index can be obtained using this approximate relation.

Therefore, the probability of failure will be very high in the case of column with high eccentricity and this fact is very important in evaluating the structure.

6.8.2 Effect of Major Limit State Variables

There are different variables that are included in the equation of the limit state. These variables are summarized in Table 6.1. The study in this section is focused on the main variables of limit state equation which affects the probability of failure and hence the reliability index value as well. In each run of the Monte-Carlo Simulation program, for each case of column dimensions and orientations, the Crystal ball program can perform a sensitivity analysis for the problem in addition to the determination of failure probability. From this, we can observe that the main variable that affects the probability of failure is the concrete strength, dead load, and the yield strength of steel bars.

Therefore, this parametric study is focused on these major variables, by using different types of concrete quality which will present different values of mean and variation in concrete strength. In the case of dead load, the value of mean and CoV depends on the quality of construction. The steel bar yield strength and its area are also variables depending on the manufacture itself.

The simulation program runs for each case and gets the relation between the reliability index and the CoV for each variable.

As a result from this study it provides us a sense in case of evaluating the building to be considered that the corner, exterior and interior columns are different in its probability of failure and also the column under eccentricity.

6.8.2.1 Effect of Concrete Strength

The strength of a concrete structure will differ somewhat from the strength of the same concrete in a control specimen for several reasons (see Chapter 3).

The mean value of concrete compressive strength in structure is calculated by using Equation 2.1.

According to MacGregor (1976), the CoV of the concrete strength in structure, V_{st}, is calculated by combining the control specimen CoV, V_c, and the CoV of the ratio between the strength in structure and that in the control specimen, V_{sr}, which is taken equal to 0.10.

Therefore, the CoV of concrete strength in structure may be written in terms of V_c and V_{sr} as follows:

$$V_{st} = \sqrt{V_{sr}^2 + V_c^2} \qquad (6.26)$$

Reliability-Based Design for Structures

In general, the CoV reflects the quality control of the concrete. Therefore, to discuss the effect of concrete strength in reliability index, a different grade of concrete quality control and corresponding different CoVs is taken into account.

Different combinations of V_c and V_{sr} are considered. These combinations yield different values for the CoV of structure concrete strength as shown in Table 6.6.

The first case, as shown in the table, is an ideal case where the strength of the concrete structure has no variation.

In the second case, it is assumed that $V_c = 0.15$ while $V_{sr} = 0.0$, leading to a variation of 0.15 for the strength of concrete structure. In the remaining cases, it is assumed that V_{sr} has a constant value of 0.1, while V_c has values in the range of 0–0.225, which yields different values of V_{st}, as shown in Table 6.6.

The reliability indices corresponding to the different values of concrete strength CoVs are represented in Table 6.6.

The reliability indices corresponding to the variation of control specimen are plotted in Figure 6.6.

It is observed that the column is more reliable ($\beta = 6.8$) when the strength of the concrete structure is deterministic (i.e., $V_{st} = 0.0$) as in the ideal case, while the column has less reliability index ($\beta = 5.6$) when the variation exists only on the concrete strength of control specimen.

In the case of concrete specimen CoV equal to 0.15, the reliability index increases by about 10% if there is no variation between the concrete control specimen and that in the structure.

Figure 6.7 shows that the reliability index of column is gradually decreased with increasing the variation of the concrete specimens.

Using a regression analysis method, one can see the relation between the reliability index and concrete control specimen CoV, V_c:

$$\beta = 5.5402 - 5.9637 V_c - 19.143 V_c^2 \qquad (6.27)$$

An approximate value of safety index can be obtained using this approximate relation.

TABLE 6.6
Relation between Concrete Strength Variation and Reliability Index

Case No.	Mean Value kg/cm², F_{cu}	Structure/Specimen Variation V_{sr}	Specimens Control Variation V_c	Structure Concrete Variation V_{st}	Reliability Index β
1	200.0	0.0	0.000	0.000	6.8
2	200.0	0.0	0.150	0.150	4.6
3	197.0	0.10	0.000	0.100	5.5
4	197.0	0.10	0.0375	0.1068	5.3
5	197.0	0.10	0.075	0.125	5.1
6	197.0	0.10	0.1125	0.150	4.6
7	197.0	0.10	0.150	0.180	4.1
8	197.0	0.10	0.225	0.246	3.3

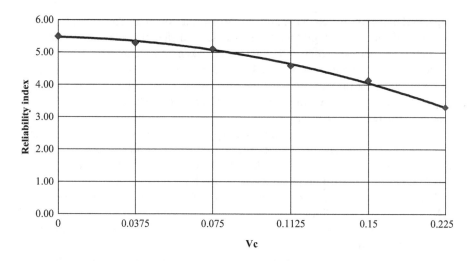

FIGURE 6.7 Relation between concrete in specimen strength variation and reliability index.

Consequently, the probability of failure is higher due to poor quality of the concretes that cannot be mitigated for selecting the materials or the project quality control system.

6.8.2.2 Effect of Dead Load

Dead loads are gravity loads resulting from the weight of the structure itself and from the weight of functional elements permanently attached to the structure.

The bias factor, λ, which is the ratio of mean to nominal value, is equal to 1.03–1.05, with CoV, V_D, equal to (0.08–0.10) (Ellingwood, 1980; Nowak, 1995).

In this section, the effect of dead load bias factors and corresponding CoV were assumed to have different values of bias factor ranging from 1.00 to 1.06.

The bias factors and corresponding CoV are shown in Table 6.7.

The reliability index calculated for each case is plotted in Figure 6.8. From this figure, one can observe that the reliability index decreases gradually with increasing the dead load bias factor.

TABLE 6.7
Relation between Dead Load Variation and Probability of Failure

Case No.	Bias Factor λ	Coefficient of Variation, V_D	Reliability Index β	Probability of Failure P_f
1	1.00	0.00	5.066	1.02×10^{-5}
2	1.01	0.06	4.694	2.9×10^{-5}
3	1.02	0.07	4.548	5.26×10^{-5}
4	1.03	0.08	4.433	8.83×10^{-5}
5	1.04	0.09	4.33	1.31×10^{-4}
6	1.05	0.1	4.135	2.85×10^{-4}
7	1.06	0.11	4.079	0.000363

Reliability-Based Design for Structures

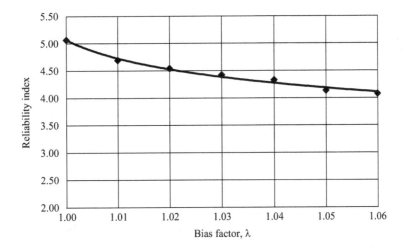

FIGURE 6.8 Relation between bias factor and reliability index.

From Table 6.7, one can conclude that the reliability index decreases, from no variation, by about 24% from that of a higher bias factor.

Using a regression analysis method, the relation between the reliability index and the dead load bias factor may be approximated by the following equation:

$$\beta = 5.0704 - 0.4939 \ln(\lambda) \qquad (6.28)$$

An approximate value of safety index can be obtained using this approximate relation.

6.8.2.3 Effect of Steel Strength

The quality control of steel bars manufacturing is reflecting on the CoV of steel strength.

To study the effect of steel strength variation on the reliability index of reinforced concrete columns, assume different values of steel strength CoV as presented in Table 6.8.

From this table, one can notice that the reliability index decreases as the CoV of the steel strength increases, but there is a slight difference of reliability index in case of no variation of steel strength and in the case of higher variation.

6.9 RELIABILITY OF FLEXURAL MEMBER

In the flexural member, the reliability analysis will be different as it will be two different limit state equations due to flexural failure and shear failure.

The prediction of the section strength is as follows:

$$R_n = A_s f_y d \left(1 - \frac{A_s f_y}{1.7 f'_c b d}\right) \qquad (6.29)$$

TABLE 6.8
Relation between Steel Strength Variation and Probability of Failure

Case No.	Coefficient of Variation, V_s	Reliability Index β	Probability of Failure P_f
1	0.00	4.18	1.43E–05
2	0.01	4.14	1.7E–05
3	0.04	4.14	1.71E–05
4	0.07	4.14	1.75E–05
5	0.10	4.12	1.9E–05
6	0.13	4.12	1.94E–05

This is used by ACI to express the ultimate bending capacity of under reinforced concrete sections, where A_s is the nominal tension steel area, f_y is the specified yield stress, d is the depth of beam, and f_c is the specified 28-day cylinder strength and b presents the nominal width of the beam.

Based on Ravindra et al. (1974), the bending resistance may be expressed as an idealization in the following product form:

$$R = \mu_R MFP \tag{6.30}$$

where

M is the random factors reflecting the uncertainty in material properties,
F is the random factors reflecting the uncertainty in fabrication, and
P is the random factors reflecting the uncertainty in simplification in strength analysis,

Accepting the validity of ultimate strength analysis, the true bending resistance may be thought of as the product of stochastic yield force and stochastic effective depth. These can be identified as M and F, respectively. The uncertainties due to the assumptions of stress block, neglect of creep, etc., are identified by the factor P.

The mean resistance, μ_R, may be evaluated as

$$\mu_R = \left(\frac{\mu_R}{\mu_{R'}}\right)\left(\frac{\mu_{R'}}{R_n}\right) R_n \approx \mu_{R/R'}\, \mu_{R'/R_n} R_n \tag{6.31}$$

where R' is the beam capacity at 28 days of age as observed in laboratory tests.

The ratio of the mean resistance of a beam section in service (at a time of maximum load during the design life) to the resistance of the section tested under laboratory conditions, $\mu_{R/R_n} R_{n'}$, for under-reinforced concrete beams in bending depends on many factors and must be estimated. The most important factors are probably aging, self-stress, stress redistribution mix control, and workmanship. Benjamin and Lind (1969) took this ratio to be 0.9. This value is rather to conservative since the beneficial effects of aging and stress redistribution would more than offset the quite minor influence of self-stress in the ultimate state, while the last two factors are negligible by comparison to obtain a more precise value, seven faculty members and research assistants experienced in reinforced concrete design at the university of waterloo, waterloo Ontario, Canada, were asked independently to estimate the mean of R/R'.

The estimates ranged from 0.95 to 1.05, averaging 1.01. The relationship between range and dispersion known for normally distributed varieties suggests that the individual estimate has a standard deviation of approx. 0.05. The value, $\mu_{R/R'} = 1.01$, will be used in this example; its estimated CoV is $0.05/\sqrt{1} = 0.02.vp$.

Finally, the mean value of observed to predicted beam capacities is $\mu_{R'/R} = 1.14$, as calculated by Sexsmith (1967) on the basis of 109 under reinforced beam tests reported in the literature. This gives $\mu_R = (1.01)(1.14)R_n = 1.15$.

The CoV of the resistance, V_R', is approximately

$$V_R \approx \sqrt{V_M^2 + V_P^2 + V_F^2} \tag{6.32}$$

The CoV of steel force is taken to be 0.09, based on extensive mill test data (Ellingwood and Ang 1972).

This CoV varies with the size of reinforcement; however, at the time of design, the designer may not know what size will be used as several different combinations may yield the same mean steel area. Thus, a constant value of V_M equal to 0.09 is used. The uncertainty in fabrication, V_F, is assumed to include the uncertainties in the effective depth, d, in the steal area, A_s, and due to different sizes of steel bars.

An analysis of available data performed by Johnson and Waries (1969, 1971) suggested a CoV value for the effective depth, d, of 0.07. Based on ASTM acceptance criteria on bar sizes, the CoV of steel area is estimated as 0.02. The CoV due to use of different sizes of bar is taken to be 0.04. These values revealed to the fabrication CoV will be calculated as follows:

$$V_F = \sqrt{(0.07)^2 + (0.02)^2 + (0.04)^2} = 0.080$$

Tichy and Vorlicek (1972) mentioned that the above value is conservative if the statistical size effect is taken into consideration.

Sexsmith (1967) reported that the CoV for the ratios of the observed to predicted beam capacities in the laboratory which is Vp is equal to 0.09. So V_R, as calculated from Equation 6.32, is equal to 0.15.

According to Arafa (2000), the flexural strength of the beam sections is mathematically modeled employing representative constitutive laws for concrete and reinforcement. Monte-Carlo technique is employed to simulate the behavior of the beam sections at their flexural limit states. The results of the reliability-based analysis are presented in terms of the reliability index at various levels of reinforcement. At reinforcement of 0.4 of the maximum permissible reinforcement ratio, the reliability index is about 4.0. This value drops to about 2.5 when the maximum permissible reinforcement is used. These results indicate that reliability of beam section is highly sensitive to variation in the compression and tension reinforcements even when the design safety factors are kept constant.

Estimation of the statistical characteristics of the beam flexural strength is an essential step toward reliability analysis. This step can be achieved through several steps explained as follows:

1. Select an appropriate constitutive law for concrete and reinforcement. In this study, the nonlinear model suggested by Hognestad et al. (1955) is employed. The descending part is presented by a straight line as shown in Figure 6.1. The stress–strain curve for reinforcement can be defined by the following three stages: (a) the linear stage with well-defined modulus of elasticity up to the yielding stress, (b) the yielding stage where the strain increases with constant stress, and (c) the stage of strain hardening which ends with the fracture of the reinforcement as shown in Figure 6.9
2. Define all possible flexural modes of failure of the beam sections. Beam sections at their flexural limit state may fail in one of two main modes depending on the section tension reinforcement ratio. The first mode of failure is the ductile failure in which the tensile strain in the reinforcement exceeds its yielding strain as the concrete compression strain reaches its ultimate strain, ε_{cu}. The second type of section failure is the brittle mode of failure in which the tensile strain of the reinforcement is less than the yield strain as the compressive strain in concrete reaches the ultimate strain.
3. Prepare an appropriate algorithm for performing the sectional analysis, defining the mode of flexural failure, and computing the flexural strength. Strength evaluation of a beam section can be achieved by several steps. The first step is to determine the depth of the neutral axis at the limit state, x_u, from the section equilibrium. The second step is to compute the strain in the tension reinforcement, ε_s, from strain compatibility assuming a perfect bond between concrete and reinforcement. This type of failure can be identified depending on ε_s. If $\varepsilon_s < \varepsilon_y$, the failure mode is brittle otherwise it is ductile mode of failure. The flexural strength of beam section can be determined by computing the moment of the tension and compression forces about any point in the section, as shown in Figure 6.10.
4. Estimate the statistical characteristics of the concrete compressive strength, the reinforcement yield strength, and the sectional dimensions. Before starting the sensitivity analysis, it is necessary to identify the practical

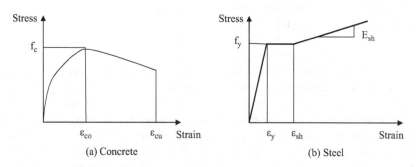

FIGURE 6.9 Constitutive models of concrete and steel.

Reliability-Based Design for Structures

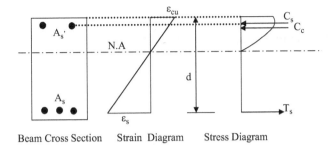

FIGURE 6.10 Strain and stress diagrams of the beam cross-section.

ranges of the statistics of the compressive strength of normal weight concrete depending on its nominal value and the level of quality control during the concrete production, casting and curing and those for the reinforcement yield strength depending on its method of production.

5. Prepare an appropriate algorithm to determine the basic statistics of flexural strength of the beam sections and their distribution function. In this study, the range of λ_c is taken between 0.8 and 1.2, whereas the range of V_c is between 20% and 40%. The range of λ_s is taken between 1.0 and 1.4, whereas V_s is taken between 4% and 12% (Arafah, 1997). The CoV of the depth of tension reinforcement, d, and compression reinforcement, d', are taken as 2% and 20%, respectively (Arafah et al., 1991).
6. Select an appropriate simulation technique for random generation of the basic strength parameters and perform the simulation process. Monte-Carlo technique is employed in this study. The flexural strength results obtained from the simulation process are plotted on the normal probability papers, NPP, and the λ_R and V_R are computed as follows:

$$\beta = \frac{\lambda_R R_n - \lambda_L L_n}{\sqrt{(\lambda_R R_n V_R)^2 + (\lambda_L L_n V_L)^2}}$$

where R_n and L_n are the nominal values of strength and load effect, respectively. The results are presented in Table 6.1. The ratio of the nominal strength to nominal load effect is kept constant in the reliability analysis, i.e., $R_n/L_n = 1.5$. The load effect λ_Q and V_Q are assumed to be 0.8 and 0.2, respectively. Reliability index is determined for various values of tension reinforcement.

The results are listed in Table 6.9 and are plotted in Figure 6.3. The results indicate that section reliability is highly sensitive to the compression and tension reinforcements even with constant safety factor. For singly reinforced sections, β dropped from 5.08 to 2.68 by increasing ρ/ρ_b from 0.3 to 0.75. The compression reinforcement improves the section reliability especially when $\rho/\rho_b > 0.5$. At $\rho/\rho_b = 0.75$, β increased from 2.68 to 3.65 by increasing ρ' from 0.0% to 10% of ρ_b.

The results indicate that reliability of reinforced concrete beam section is highly sensitive to variation in the compression and tension reinforcements even when

TABLE 6.9
Results of Reliability-based Sensitivity Analysis[a]

	ρ/ρ_b	λ_R	V_R	β
Case 1	0.3	1.23	0.07	5.08
$\rho'/\rho_b = 0.00$	0.4	1.18	0.078	4.59
	0.5	1.12	0.093	3.94
	0.6	1.07	0.108	3.41
	0.75	0.98	0.131	2.68
Case 2	0.3	1.25	0.068	5.25
$\rho'/\rho_b = 0.05$	0.4	1.21	0.073	4.89
	0.5	1.16	0.087	4.27
	0.6	1.09	0.096	3.73
	0.75	1.04	0.114	3.18
Case 3	0.3	1.28	0.067	5.46
$\rho'/\rho_b = 0.10$	0.4	1.24	0.070	5.14
	0.5	1.22	0.090	4.49
	0.6	1.17	0.105	3.91
	0.75	1.08	0.100	3.60

[a] Note that $R_n/L_n = 0.5$, $\lambda_L = 0.8$, and $V_Q = 0.2$.

the design safety factors are kept constant. High reliability levels are observed for low-tension reinforcement ratios. At reinforcement of 0.4 of the maximum permissible reinforcement ratio, the reliability index is about 4.0. This value drops to about 2.5 when the maximum permissible reinforcement is used. The compression reinforcement increases the section ductility and reliability. Compression reinforcement is highly recommended in beam sections.

6.10 SEISMIC RELIABILITY ANALYSIS OF STRUCTURES

Reliability analysis of structures implies estimation of the limit state probabilities of a structure under the environmental load affecting the building during its service life. There is a synonymous nomenclature, called safety, which is used to indicate reliability. Similarly, risk analysis and reliability analysis of structures are simultaneously used in many publications to express their probabilities of failure. However, they are not actually one and the same thing. Risk analysis of structures is an extension of the reliability analysis as it is the probability of failure multiplied by the consequence.

Typically in the seismic risk analysis of structures, the limit state probability of the structure which will be obtained from the reliability analysis is integrated with the seismic risk of the site. An associated term that is used in connection with the seismic reliability or risk analysis of structures is the fragility analysis. Fragility analysis is aimed at finding the probability of failure of structures for different levels of PGA at the site and is closer to the seismic risk analysis of structures. Despite

Reliability-Based Design for Structures

these finer distinctions, seismic risk, reliability, safety, and fragility analysis of structures are used loosely in the literature to denote the seismic probability of failure of structures, failure being defined by different limit state conditions.

The most important aspect of the reliability analysis is the consideration of uncertainties that make structures vulnerable to failure for a predefined limit state. Accuracy of the reliability analysis depends upon how accurately all the uncertainties are accounted for in the analysis. Firstly, it is practically impossible to identify all uncertainties; however, important ones can be identified. Secondly, and most importantly, methods for modeling and analyzing them are not easy and some amount of uncertainty always remains associated with their modeling. Finally, analytical formulation of the limit state surface and integration of the probability density function within the domain of interest are complex resulting in various approximations. As a result, varying degrees of simplifications are made in the reliability analysis leading to the development of different reliability methods. Therefore, it is not possible to obtain the exact probability of failure of a structure for any event except for very simple ones.

In this chapter, seismic reliability analysis of structures is briefly discussed. As the subject is vast and considerable research has taken place in this area, it is not possible to cover the entire subject in one small chapter. Only some of the fundamental concepts of seismic reliability of structures, and a few simpler Generally there are three types of uncertainties, which are dominant in seismic reliability analysis, namely:

a. randomness and variability of excitation;
b. statistical uncertainty, which arises due to estimation of parameters describing statistical models; and
c. model uncertainty, which arises due to imperfection of mathematical modeling of the complex physical phenomena.

In addition to these uncertainties, there are some others which result from simplification of the problem at hand, for example, nonlinear analysis may be replaced by equivalent linear analysis, continuums may be represented by a discrete model with limited degrees of freedom, and so on.

In the main, the uncertainty arising due to (a) is irreducible but those arising due to (b) and (c) can be reduced. For example, collection of more data or samples helps in providing better statistical parameters.

Likewise, use of a more refined model may reduce the uncertainty due to (c). Other uncertainties, as mentioned above, may be reduced by performing more rigorous analysis with more sophisticated models of structures.

In the seismic reliability analysis of structures, uncertainties of earthquake or ground intensity parameters are considered to affect the reliability estimates significantly and to assume more importance over other uncertainties. As a result, seismic reliability analysis of structures is found to be mostly carried out by considering the randomness of ground motion, the uncertainties inherent in the occurrence of an earthquake and in defining its different intensity parameters. However, many problems have also been solved in which uncertainties of material behavior and modeling of the physical phenomenon have also been included along with the uncertainties of ground motion and earthquakes. The complexity of the analysis increases manifold as more sources of uncertainties are included.

In the seismic reliability analysis of structures, the randomness of ground motion, the uncertainties of earthquake, and its intensity parameters are included in several ways by considering some or all of the following elements of seismicity:

i. Power spectral probability density function (PSDF) of ground motion.
ii. Risk-consistent or uniform-hazard response spectrum.
iii. Model of the occurrence of an earthquake.
iv. Attenuation laws.
v. Probability density functions of the magnitude of an earthquake, epicentral distance, and sources of an earthquake.
vi. Hazard curve.
vii. Empirical relationships to describe the response spectrum ordinates as a function of magnitude and epicentral distance.

The probabilistic models that are widely used to describe the distributions of different uncertain parameters are uniform distribution, extreme value distribution, log normal distribution, and Poisson distribution.

The procedures for performing the reliability analysis vary with the selection of the above elements. When material and other uncertainties are introduced, the procedures for the analysis may significantly differ, for example, stochastic finite-element analysis for and loading may be used. However, material and other uncertainties may also be included by simple procedures in an approximate manner. In fact, various levels of approximations are often used to simplify the reliability analysis procedure consistent with the desired accuracy.

The failure of the Haitian palace is due to earthquake that was happened on 2010 as shown in Figure 6.11. The Haiti earthquake was a catastrophic magnitude 7.0 M_w earthquake, with an epicenter near the town of Léogâne, approximately 25 km (16 miles) west of Port-au-Prince, Haiti's capital.

FIGURE 6.11 Collapse of Haiti palace during earthquake.

Reliability-Based Design for Structures

Seismic reliability or risk analysis of structures can be performed with different degrees of complexity as outlined previously. It is practically impossible to consider all uncertainties in one analysis and the summary of uncertainties that could be considered in the analysis are as follows (Datta 2010):

1. Uncertainty of earthquake
    a. Uncertainty in point of occurrence along the fault
    b. Uncertainty of earthquake size
    c. Uncertainty in time
2. Uncertainty associated with seismic hazard estimates
    a. Uncertainty of attenuation laws
    b. Uncertainty of empirical laws used
    c. Uncertainty of site amplification
3. Uncertainty of ground motion input
    a. Risk consistent spectrum
    b. Hazard curve
    c. Design spectrum with a return period
    d. PSDF of ground motions and envelope function
4. Uncertainty of the structures modeling
    a. Mass modeling
    b. Stiffness modeling
    c. Damping modeling
5. Uncertainty of analysis
    a. Random vibration analysis
    b. Nonlinear dynamic analysis being replaced by equivalent linear analysis
    c. Selection of failure mechanism
6. Uncertainty of material property
    a. Elastic constant
    b. Working stress of limit state stress
    c. Ductility

Furthermore, the estimated probability of failure obtained by any reliability analysis technique cannot be accurate because of the approximations involved in each method. Thus, the calculated seismic reliability is at best a good estimate of the actual reliability of structures against the failure event. In view of this, many amplified seismic reliability analyses have been proposed by various researchers. They are useful in obtaining an estimate of the seismic reliability of structures considering some (but not all) of the uncertainties at a time. Some of them are described here. They include (i) reliability analysis of structures considering uncertainty of ground inputs only; (ii) reliability analysis of structures using seismic risk parameters of the site; (iii) threshold crossing reliability analysis of structures for deterministic time history of ground motion; (iv) first passage reliability analysis of structures for random ground motions; (v) reliability analysis using damage probability matrix; and (vi) simplified probabilistic risk analysis of structures.

Reliability Analysis of Structures Considering Uncertainty of Ground Input. The simplest analysis that can be performed is that of finding the seismic reliability of a structure using a risk consistent spectrum (or uniform hazard spectrum) that considers the uncertainty of ground motion input only. The probability of failure of the structure designed with this spectrum is the exceedance probability of the response spectrum ordinate being used for the design. If a uniform hazard spectrum or risk consistent response spectrum is constructed using seismic hazard analysis, then it is implicit that the analysis also considers uncertainties (2) and (1). Thus, if a rigorous seismic hazard analysis has been performed for a site and the resulting seismic input in the form of a risk consistent spectrum is used in the seismic analysis of the structures, then the uncertainties indicated in (1), (2), and (3a) are mostly included in the reliability estimate. Note that the estimated probability of failure does not include other uncertainties as discussed above and that the failure of the structure is assumed to take place under the load, determined from the spectrum.

6.11 EXAMPLE

Consider a multi-story frame as studied by Datta (2010) as shown in Figure 6.12. Three alternative designs of the frame are made with the help of three risk consistent response spectrums, which have 10%, 5%, and 2% probabilities of exceedance in 50 years, respectively. It is assumed that the frame collapses under the earthquake loads obtained by the response spectrums. Determine the probabilities of failure of the frame.

As the frame collapses under the lateral loads determined using the response spectra, the probabilities of exceed of the response spectrum ordinates are the probabilities of failure of the frame.

The probability of exceed of at least one or more in t years, for the response spectrum ordinates, is given by the commonly used equation:

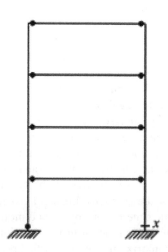

FIGURE 6.12 Probable mechanism of failure under seismic load.

Reliability-Based Design for Structures

$$P_e = 1 - \left(1 - \frac{1}{\tau}\right)^t$$

where P_e is the probability of exceed in t years and t is the average recurrence interval. Using the above equation, the average recurrence intervals are calculated as follows:

- 475 years for 10% probability of exceed in 50 years
- 975 years for 5% probability of exceed in 50 years
- 2475 years for 2% probability of exceed in 50 years

The corresponding probabilities of the failure are as follows: $2:10 \times 10^{-3}$, $1:025 \times 10^{-3}$, and $4:040 \times 10^{-4}$, respectively.

6.12 STEEL AND OFFSHORE STRUCTURE RELIABILITY

For the steel structure, the reliability of analysis will be the same procedure and equation except the main reason of reducing reliability of the existence of corrosion. In the case of offshore structure, there are a lot of studies about its reliability until now as there are a lot of uncertainties on the wave that affect the offshore structure.

The other important factor to define the building reliability is the redundancy of the building. For example, it is found by destroying the prototype of offshore jacket that the X-bracing has a higher redundancy more than the k-bracing system. So the probability of failure of k-bracing system is higher than that in case of X-bracing. To understand the redundancy, let us see the following case as in Figures 6.13 and 6.14.

If you need to design the beam and you receive the following two solutions, the first is a structural system to be fixed in the two ends and the second structural system is hinge in the two ends. Which system will you choose? Give yourself 5 min to think.

There are many factors that control your decision in selecting the suitable system. The following is the advantages and disadvantages for these two systems.

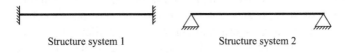

FIGURE 6.13 Comparison between structural redundancy.

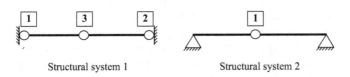

FIGURE 6.14 Comparison between structural redundancy after failure.

Structural system 1:

- The beam cross-section will be smaller.
- The connections will be big and complicated as has a shearing force and moment on it.
- It is reasonable from architectural point of view.
- The construction is complicated in connection.

Structural system 2:

- The beam cross-section will be big.
- The connection will be small as it designs to shear force only.
- It is easy to construct it as the connection is simple.

So by discussing the above two systems, if you are afraid from construction onsite, or maybe you will carry out the construction with the inhouse engineers and workers to avoid any problems, you will go toward the simple beam option. This type of thinking is always happened as we always forget the maintenance point of view. From the below figure, you will find the steps of collapse failure.

The structural system 2, which is the simple beam, assumes that the load is increased gradually and the beam can accommodate the load until a plastic hinge in the middle of the beam at the point of maximum bending moment. Then collapse will occur.

On the other hand, for the structural system 1, which is fixed at the two ends, when the load is increased gradually, the weaker of the left or right connection will be fail first. As shown in the figure, the plastic hinge will form on the left connection and increase the load; the other connection will be a plastic hinge, so it is now working as simple beam. Then by increasing the load, plastic hinge 3 will be formulated and then collapse failure will occur.

From the above discussion, one can find that structural system 1 will take more time to collapse as it fails after three stages. Structural system 2 fails in first stage, so structural system 1 is more redundant than system 2.

Moreover when you make a comparison between different reinforced concrete members such as slabs, beams, columns, and cantilevers, one can obtain that there are some members that most critical than other. As cantilever is most critical structure member as any defect on it will produce a high deflection and then failure, in addition to that the column when column is fail the load shall be distributed to other columns until failure of the whole building. Therefore, the column is defined that it has a low redundancy and very critical as any failure on it will fail the whole building however the cantilever failure it will be a member failure only so we can go through the consequences from this approach.

In the case of beam and slab, when you design a slab, assume simple beam and calculate the maximum moment in the middle point (in case of simple beam) and then design the concrete slab by choosing the slab thickness and the steel reinforcement based on this point. In actual the selected steel reinforcement will be distributed along the whole span during constriction, so theoretically by increasing the load the failure will be in a point only but actually the surrounding area will carry part of this load so the redundancy of the slab is very high.

In some researches, the reinforced concrete slab can accommodate load twice the design load.

For the structure as a whole, we can use Pushover analysis to obtain the redundancy of the structure. This analysis is nonlinear analysis and now is for any structure analysis software in the market. From this analysis, one can obtain how much more load can the structure carry than the design load until failure. In addition, we can find that the location of the first plastic hinge will formulate and thus one can obtain the critical member in your structure.

BIBLIOGRAPHY

ACI Committee 318, Building code requirements for reinforced concrete and commentary, ACI 318M-95, American Concrete Institute, Farmington Hill, 1995.

ACI-318, Building code requirement for reinforced concrete (ACI-95) and commentary-(ACI318–95R), American Concrete Institute, Detroit, MI, 1995.

Ahmed, M.A., Minimum cost design of reinforced concrete short columns, M.Sc. thesis, Faculty of Engineering, Cairo University, Cairo, 1985.

Ahmed, M.A., Reliability analysis of ductile plan structures, Ph.D. Thesis, Faculty of Engineering, Cairo University, Giza, 1990.

Ang, A.H., Tang, W.H., 1984, *Probability Concepts in Engineering Planning and Design.* Vol. II, Decision, Risk and Reliability, John Wiley & sons, Inc., Hoboken, NJ.

Arafah, A., 1997, Statistics for concrete and steel quality in Saudi Arabia, *Magazine of Concrete Research*, Vol. 49, No. 180, pp. 185–194.

Arafah, A., Reliability of reinforced concrete beam section as affected by their reinforcement ratio, *8th ASCE Specialty Conference on Probabilistic Mechanics and Structural Reliability,* PMC2000–094, 2000.

Arafah, A.M., Integration of human errors in structural reliability models, Ph.D. Thesis, University of Michigan, Ann Arbor, MI, 1986.

Arafah, A., Al-Zaid, R., Al-Haddad, M., AL-Tayeb, A., Al-Sulimani, G., and Wafa, F., 1991, *ASCE*, Vol. 107, No. 6, pp. 1133–1153.

Arafah, A., Al-Zaid, R., Al-Haddad, M., AL-Tayeb, A., Al-Sulimani, G., and Wafa, F., 1991. Development of a solid foundation for a national reinforced concrete design building code, Final Report, KACST Project No. AT-9-34, Riyadh, Saudi Arabia.

Benjamin, J.R., and Lind, N.C., 1969, A probabilistic basis for a deterministic code, *Journal of the American Concrete Institute*, Vol. 66, No. 11, pp. 857–865.

Box, G.E.P., and Tiao, G.C., 1973, *Bayesian Inference in Statical Analysis*, Addison-Wesley, Reading, MA.

Cornell, C.A., 1967, Bounds on the reliability of structural systems, *Journal of the Structural Division*, Vol. 93, No. 1, pp. 171–200.

Datta, K.A., 2010, *Seismic Analysis of Structures*, Wiley, Hoboken, NJ.

Der Kiureghian, A.D., 1981, Seismic risk analysis of structural systems. *Journal of Engineering Mechanics, ASCE*, Vol. 107, No. 6, pp. 1133–1153.

Der Kiureghian, A.D., 1996, Structural reliability methods for seismic safety assessment: A review, *Engineering Structures*, Vol. 18, No. 6, pp. 412–424.

Ellingwood, B.R., and Ang, A.H, A probabilistic study of safety criteria for design, structural research series No.387, University of Illinois, Urbana, 1972.

Ellingwood, B., MacGregor, J.G., Galambos, T.V., and Cornell, C.A., 1982, Probability based load criteria- load factors and load combinations, *Journal of Structural Division, ASCE*, Vol. 108, No. 5, pp. 978–997.

El-Reedy, M.A., Ahmed, M.A., and Khalil, A.B., Reliability analysis of reinforced concrete columns, Ph.D. Thesis, Faculty of Engineering, Cairo University, Giza, 2000.

Frangopol, D.A., Spacopne, E., Milner, D.M., 2001, New light on performance of short and slender reinforced concrete columns under random loads, *Engineering Structures*, Vol. 23, pp. 147–157.

Hognestad, E., Hanson, N.W., and McHenry, D., 1955, Concrete stress distribution in ultimate strength design, *ACI Journal, Proceedings*, Vol. 52, No. 4, pp. 455–479.

Johansson, A., and Warris, B., Deviations in the location of reinforcement, *Proceedings No.40*, Swedish cement and concrete institute, Royal institute of technology, Stockholm, Sweden, 1969.

Johnson, A.I., Strength, safety and economical dimensions of structures, bulletin No. 12, Royal institute of technology division of building studies and structural engineering, Stockholm, Sweden, 1953; document D7, national Swedish building research, 1971.

MacGregor, J.G., 1976, *Safety and Limit states Design For Reinforced Concrete*, The University of Alberta, Edmonton, AB.

Nowak, A.S., 1979, Effect of human errors on structural safety, *Journal of the American Concrete*, Vol. 76, No. 9, pp. 959–972.

Nowak, A.S., 1995, Calibration of LRFD bridge design code, *ASCE Journal of Structural Engineering*, Vol. 121, No. 8 pp. 1245-1251.

Sexsmith, R.G., Reliability analysis of concrete structure, Report No. 83, Department of Civil Engineering, Stanford University, Stanford, CA, 1967.

Tichy, M., and Vorlicek, M., 1972, *Statistical Theory of Concrete Structures with Reference to Ultimate Design*, Academia, Prague.

User Manual, Crystal ball program version 3.0, Decision Engineering, 1996.

7 Reliability of Concrete Structure Exposed to Corrosion

7.1 INTRODUCTION

Most of the structure deterioration for concrete, steel, and offshore structure occurs due to corrosion. This corrosion can cause a complete structure failure. This occurs clearly in LPG sphere tank which causes a catastrophic failure with high consequence. The LPG sphere legs are covered by fire-rated materials in most case by concrete with steel mesh. The normal safety precaution procedures to test the firefighting equipment every 2 weeks. The firefighting water will cause corrosion between the steel leg and concrete, but this reduction of the thickness cannot be checked visually, and you shall remove part of the cover to check it. This complete failure occurs due to corrosion of the leg and concurrent in same time fill the tank with water to the hydrostatic test. This chapter illustrates the causes of corrosion of the steel bar reinforcement and defines the structure probability of failure depending on the structure condition. The concrete structure reliability depends on the structure design methodology. This chapter describes a better option of designing to increase the structure life time in the case of exposed the structure to corrosion.

Most of the design studies in reinforced concrete literature assume that the durability of reinforced concrete structures can be taken for granted. However, many reinforced concrete structures are exposed during their life to environmental stress (e.g., corrosion, expansive aggregate reactions), which attack the concrete and/or steel reinforcement (Kilareski, 1980; Cady and Weyers, 1994; West and Hime, 1985, Takewaka and Matsumoto, 1988; Mori and Ellingwood, 1994a; Lin, 1995; Thoft-Christensen, 1995).

In this chapter, a reliability of the reinforced concrete structure under the effect of corrosion attack and increasing the concrete strength with time is studied.

Effect of age on concrete strength is discussed in Section 7.2.

The corrosion of reinforced steel bars as concrete characteristic is reviewed in Section 7.3.

In Section 7.4, a parametric study is carried out on the reliability of reinforced concrete columns taking into account the corrosion. The effect of age on the reliability of reinforced concrete column for wet and dry conditions is presented in Section 7.4.1, the effect of longitudinal steel ratio of reinforced concrete columns on the reliability is discussed in Section 7.4.2, and the effect of corrosion rate on the reliability of reinforced concrete columns is discussed in Section 7.4.3, respectively.

In Section 7.4.4, the effect of initial time of corrosion on the reliability of reinforce concrete structure members is presented. The effect of corrosion on the reliability of eccentrically loaded columns is studied in Section 7.4.5.

7.2 EFFECT OF AGE ON STRENGTH OF CONCRETE

Many researches were done to predict the concrete strength after 28 days. As in the majority of cases, the tests are made at the age of 28 days when the strength of concrete is considerably lower than its long-term strength.

Different methods have been suggested to predict the concrete strength with age. Moreover, there are different recommendations in different codes for predicting concrete strength with age.

There are many the researches that studied the gain in concrete strength with age, but the environmental conditions that surround each test and research should be considered. In addition, there are different codes that recommend equations that will predict the gain in concrete strength with age.

7.2.1 Researchers' Suggestions

The effect of time on the strength of concrete is given in the following sections. The variation of concrete strength with age in wetted and dry conditions was studied by Baykof and Syglof (1976). Baykof found that, at dry condition after 1 year, there is no increase in concrete strength as shown in Figure 7.1. On the other hand, the strength of specimens stored in wet environments (at 15°C) is considerably higher, as shown in Figure 7.1.

There is another unique study performed to investigate the change in the concrete properties after 50 years. This study was performed by Washa et al. (1989), as they have tested concrete specimens stored in special environmental conditions to predict the concrete strength with age. The specimens were moist cured for 28 days before

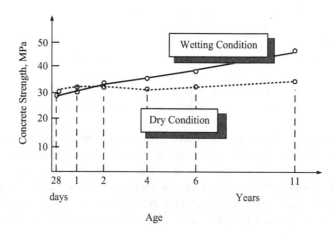

FIGURE 7.1 Variation of concrete strength with time from Baykof (1976) years.

Reliability of Concrete Structure

placement outdoors on leveled ground in an uncovered, open location. Thermocouple data indicate that the outdoor compressive cylinders were subjected to about 25 cycles of freezing and thawing each winter. The relative humidity in Madison normally varies from 65% to 100% with an average of 75%. The annual precipitation including snowfall is about 32 inch. Air temperatures usually range between 25°F and 90°F (32°C and 35°C).

The average compressive strength with time is as shown in Figure 7.2.

Washa et al. (1989) concluded that the compressive strength of concrete cylinders stored outdoors for 50 years in Madison, Wisconsin (made with cement having a relatively low C_2S content and a high surface area) generally increased as logarithm of the age for about 10 years. After 10 years, the compressive strength decreased or remained essentially the same.

MacGregor et al. (1983) used the same study and formulated the equation of the relation between compressive strength and age, which is obvious (in the case of 28-day-old concrete) that specified compressive strength of concrete equals 281.5 kg/cm² (27.6 MPa). The corresponding mean of compressive strength at 28 days is 292.7 kg/cm² (28.7 MPa).

$$f_c(t) = \begin{cases} 15.85 + 4.03\ln(t) & \text{MPa} \quad t < 10 \text{ years} \\ 48.9 & \text{MPa} \quad t \geq 10 \text{ years} \end{cases} \quad (7.1)$$

where $f_c(t)$ is the concrete compressive strength with time and t is the time in days.

The effect of maturity of concrete, based on a number of studies of the long-term strength gain of concrete, representative lower bound relationships between age and strength of various classes of concrete was obtained. These tended to show a linear relationship between strength and the logarithm of age. For lower strength concrete, 25 years strength approaching 240% of the 28 days strengths were observed (Washa

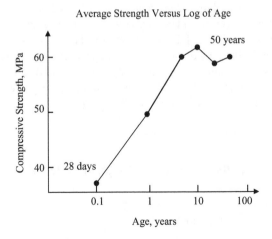

FIGURE 7.2 Variation of compressive strength with age (adopted from Washa, G.W., and K.F. Wendt. 1975, ACI J. Proc. 72 (1):20–28).

et al. 1975). For high-strength concrete, the strength of old concrete approached 125%–150% of the 28 days strength.

MacGregor (1983) assumed that there was an equal probability that a structure would be loaded to failure in each month of a 50-year lifetime starting when the structure was 28-day-old; a family of distributions of the ratio of the concrete strength at failure, f_{cf}, to the 28-day strength f_{C28} was obtained. They noticed that, for each combination of curing and concrete type, the resulting distribution of f_{cf}/f_{C28} had a negative skewness.

The effect of concrete maturing can be included in Monte-Carlo studies by multiplying the site concrete strength by a randomly selected value of f_{cf}/f_{C28} from the appropriate distribution.

Alternatively, a weighted combination of these distributions could be approximated by a normally distributed variable with a mean value of 1.25 and coefficient of variation equal to 0.07.

7.2.2 Code Recommendations

Different codes recommended different methods to predict the concrete strength for different ages. In the following sections, the gain in concrete strength with age in Egyptian code, British Code, and Indian code is discussed.

The concrete compressive strength varies with age for normal concrete at moderate temperatures (Hilal 1980). The ratio of the compressive strength 28 days to that at a given concrete age may be estimated for normal and rapid hardening Portland cement, according to ECP-98 from the values given in Table 7.1.

According to Nevil (1981), in the past, the gain in strength beyond the age of 28 days was regarded merely as contributing to an increase in the factor of safety of the structure.

Since 1957, the codes of practice for reinforced and prestressed concrete allow the gain in strength to be taken into account in the design of structures that will not be subjected to load until a large age except when no-fines concrete is used; with some lightweight aggregates, verifying tests are advisable. The values of strength given in the British code of practice CP110:1972, based on the 28 day compressive strength, are given in Table 7.2 but they do not, of course, apply when accelerators are used.

Table 7.2 shows that concrete continuously gains strength with time. The additional strength is about 20%–25% of the corresponding 28-day strength. The most of the additional strength is gained within the first year.

TABLE 7.1
Ratio of f_{C28} to that at the Age (days) at Egyptian Code 1989

Age (days)		3	7	28	90	360
Type of Portland cement	Normal	2.5	1.5	1.00	0.85	0.75
	Rapid hardening	1.80	1.30	1.00	0.90	0.85

Source: Egyptian code (1989).

TABLE 7.2
British Code of Practice CP110:1972 Factor of Increase in Compressive Strength of Concrete with Age

	Age Factor For Concrete with a 28-day Strength (MPa)		
Months	20–30	40–50	60
1	1.00	1.00	1.00
2	1.10	1.09	1.07
3	1.16	1.12	1.09
6	1.20	1.17	1.13
12	1.24	1.23	1.17

The available test results presenting the gain in concrete strength with age are discussed by Washa et al. (1989) as they tested concrete specimens at 1, 5, 10, 25, and 50 years. The analysis of this data is shown in Table 7.3 and the statistical parameters for the ratio of concrete strength at 1, 5, 10, 25, and 50 years and that at 28 days are calculated.

Baykof et al. (1976) discussed the gain in concrete strength of concrete specimens stored in wet and dry conditions with time. The factor of increase in concrete strength with age at wet and dry conditions are presented in Table 7.4.

TABLE 7.3
Analysis for the Results of Washa et al. (1989) Tests for Gain of Concrete Strength with Time

Test No.	28 days f_{c28} psi	1 year f_{c1} psi	f_{c28}/f_{c1}	5 year f_{c5} psi	f_{c28}/f_{c5}	10 years f_{c10} psi	f_{c28}/f_{c10}	25 years f_{c25} psi	f_{c28}/f_{c25}	50 years f_{c50} psi	f_{c28}/f_{c50}
1	3,990	5,195	1.30	7,510	1.88	7,865	1.97	7,585	1.90	7,650	1.92
2	5,030	6,530	1.30	8,780	1.75	8,700	1.73	8,200	1.63	9,220	1.83
3	4,285	5,805	1.35	7,400	1.73	7,855	1.83	7,840	1.83	7,610	1.78
4	4,525	6,015	1.33	7,540	1.67	8,025	1.77	7,885	1.74	7,920	1.75
5	6,185	8,195	1.32	9,015	1.46	10,465	1.69	9,850	1.59	10,400	1.68
6	4,990	6,240	1.25	7,820	1.57	7,885	1.58	8,050	1.61	8,030	1.61
7	5,150	7,395	1.44	7,940	1.54	8,525	1.66	8,070	1.57	8,820	1.71
8	6,265	8,310	1.33	10,430	1.66	10,145	1.62	9,915	1.58	10,300	1.64
9	5,130	7,325	1.43	8,860	1.73	9,015	1.76	8,250	1.61	7,950	1.55
10	5,145	6,715	1.31	7,640	1.48	7,910	1.54	6,835	1.33	7,140	1.39
11	7,050	8,695	1.23	9,995	1.42	9,515	1.35	8,985	1.27	9,250	1.31
12	5,480	7,360	1.34	8,405	1.53	8,180	1.49	7,500	1.37	7,200	1.31
Mean			1.33		1.62		1.67		1.59		1.62
St. Dev.			0.06		0.14		0.17		0.19		0.20
CoV			0.05		0.09		0.10		0.12		0.12

TABLE 7.4
Factor of Increase in Concrete Strength with Age At Different Conditions (Baykof et al., 1976)

Years	Environment	
	Dry	Wet
1mo	1.00	1.00
1	1.25	1.20
2	1.25	1.30
4	1.25	1.40
6	1.25	1.50
10	1.25	1.63

In general, the buildings are always exposed to a dry condition after construction. Therefore, the case of dry condition is taken into consideration in our study considering the factor of increase strength after 1 year by 1.25 to the strength at 28 days with CoV equal to 0.1 and this gain remains constant along the life of the building (Table 7.5).

Based on the European code EC2, It may be required to specify the concrete compressive strength, $f_{ck}(t)$, at time t for a number of stages (e.g., demolding, transfer of prestress), where

$$f_{ck}(t) = f_{cm}(t) - 8(MPa) \quad \text{for} \quad 3 < t < 28 \text{ days}$$

$$f_{ck}(t) = f_{ck} \quad \text{for} \quad t \geq 28 \text{ days}$$

More precise values should be based on tests especially for $t \leq 3$ days.

The compressive strength of concrete at an age t depends on the type of cement, temperature, and curing conditions. For a mean temperature of 20°C and curing in accordance with EN 12390, the compressive strength of concrete at various ages $f_{cm}(t)$ may be estimated as follows:

$$f_{cm}(t) = \beta_{cc}(t) f_{cm} \tag{7.2}$$

TABLE 7.5
Indian Specifications for Increasing Concrete Strength with Time

Age per Month	Age Factor
1	1.00
3	1.10
6	1.15
12	1.20

Reliability of Concrete Structure

with

$$\beta_{cc}(t) = \exp\left\{s\left[1-\left(\frac{28}{t}\right)^{0.5}\right]\right\} \qquad (7.3)$$

where
$f_{cm}(t)$ is the mean concrete compressive strength at an age of t days,
f_{cm} is the mean compressive strength at 28 days according to Table 7.1,
$\beta_{cc}(t)$ is a coefficient which depends on the age of the concrete t,
t is the age of the concrete in days,
s is a coefficient which depends on the type of cement based on EC2:
   = 0.20 for cement of strength Classes CEM 42,5 R, CEM 52,5 N, and CEM 52,5 R (Class R),
   = 0.25 for cement of strength Classes CEM 32,5 R and CEM 42,5 N (Class N),
   = 0.38 for cement of strength Classes CEM 32,5 N (Class S)

where the concrete does not conform with the specification for compressive strength at 28 days and the use of Expressions 7.2 and 7.3 is not appropriate.

This clause should not be used retrospectively to justify a non-conforming reference strength by a later increase of the strength.

The development of tensile strength with time is strongly influenced by curing and drying conditions as well as by the dimensions of the structural members. As a first approximation, it may be assumed that the tensile strength $f_{ctm}(t)$ is equal to:

$$f_{ctm}(t) = \left(\beta_{cc}(t)\right)\alpha \cdot f_{ctm} \qquad (7.4)$$

where $\beta_{cc}(t)$ follows from Expression 7.3 and
$\alpha = 1$ for $t < 28$
$\alpha = 2/3$ for $t \geq 28$. The values for f_{ctm} are given in Table 7.6.

TABLE 7.6
Strength Classes for Concrete

Properties					Concrete Strength (MPa)								
f_{ck}	12	16	20	25	30	35	40	45	50	55	60	70	80	90
$f_{ck,cube}$	15	20	25	30	37	45	50	55	60	67	75	85	95	105
f_{cm}	20	24	28	33	38	43	48	53	58	63	68	78	88	98
f_{ctm}	1.6	1.9	2.2	2.6	2.9	3.2	3.5	3.8	4.1	4.2	4.4	4.6	4.8	5.0
$F_{ctk0.05}$	1.1	1.3	1.5	1.8	2.0	2.2	2.5	2.7	2.9	3.0	3.1	3.2	3.4	3.5
$F_{ctk0.95}$	2.0	2.5	2.9	3.3	3.8	4.2	4.6	4.9	5.3	5.5	5.7	6.0	6.3	6.6
E_{cm} (GPa)	27	29	30	31	33	34	35	36	37	38	39	41	42	44
ε_1, %	1.8	1.9	2.0	2.1	2.2	2.25	2.3	2.4	2.45	2.5	2.6	2.7	2.8	2.8

7.3 CORROSION OF STEEL IN CONCRETE

This section discusses the basics of corrosion, how they apply to steel in concrete, corrosion rate, and corrosion effect on spalling of concrete.

The concrete is alkaline. Alkalinity is the opposite of acidity. Metals corrode in acids, whereas they are often protected from corrosion by alkalis.

The concrete is alkaline because it contains microscopic pores with high concentrations of soluble calcium, sodium, and potassium oxides. These oxides form hydroxides derived from the reactions between mix water and Portland cement particle which are very alkaline.

A measure of acidity and alkalinity, pH, is based on the fact that the concentration of hydrogen ions (acidity) times hydroxyl ions (alkalinity) is 10^{-14} moles/L in aqueous solution.

According to Broomfield (1997), a strong acid has pH = 1 (or less), a strong alkali has pH = 14 (or more), a neutral solution has pH = 7. Concrete has a pH of 12–13. Steel corrodes at pH 10–11.

Concrete creates a very alkaline condition within pores of the hardened cement matrix that surrounds aggregate particles and the reinforcement.

The alkaline condition leads to a "passive" layer on the steel surface. A passive layer is a dense, impenetrable film, which, if fully established and maintained, prevents further corrosion of the steel. The layer formed on steel in concrete is probably part metal oxide/hydroxide and part mineral from the cement. A true passive layer is a very dense, thin layer of oxide that leads to a very slow rate of oxidation (corrosion).

Once the passive layer breaks down, then areas of rust will start appearing on the steel surface. The chemical reactions are the same whether corrosion occurs by chloride attack or carbonation.

The corrosion is seen as a three-phase process: the first phase spans from the time of construction to the time of corrosion initiation. This phase is the diffusion of CO_2 to cause depassivation. The second phase follows until unacceptable levels of section loss have occurred. The third phase is occurred through the second phase as deterioration, which begins with cracking and spalling of concrete cover. The deterioration curve is as shown in Figure 7.3.

The three phases will be discussed in the following sections.

Figure 7.4 presents the corrosion on steel reinforcement of a balcony in which is due to a lack of maintenance. This building is on the Mediterranean sea coast (no more than 300 m), so special design considerations are needed.

7.3.1 CAUSES AND MECHANISMS OF CORROSION AND CORROSION DAMAGE

According to Broomfield (1997), the main causes of corrosion of steel in concrete are chloride attack and carbonation. These two mechanisms are unusual where they do not attack the integrity of the concrete. Instead, aggressive chemical species pass through the pores in the concrete and attack the steel.

This is unlike normal deterioration processes due to chemical attack on concrete. Other acids and aggressive ions such as sulfate destroy the integrity of the concrete

Reliability of Concrete Structure

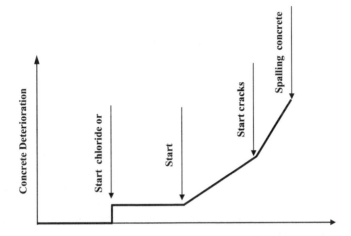

FIGURE 7.3 Sketch representing concrete structure deterioration process.

FIGURE 7.4 Corrosion of balconies.

before the steel is affected. Most forms of chemical attacks are therefore concrete problems before they are corrosion problems. Carbon dioxide and chloride ions are very unusual in penetrating the concrete without significantly damaging it. Accounts of acid rain causing corrosion of steel embedded in concrete are unsubstantiated. Only carbon dioxide and the chloride ion have been shown to attack the steel and not the concrete.

7.3.2 CARBONATION

Carbonation is the result of the interaction of carbon dioxide gas in the atmosphere with the alkaline hydroxides in the concrete. Like many other gases, carbon dioxide dissolves in water to form an acid. Unlike most other acids, the carbonic acid does not attack the cement paste, but just neutralizes the alkalis in the pore water, mainly forming calcium carbonate.

There is a lot more calcium hydroxide in the concrete pores than can be dissolved in the pore water. This helps maintain the measure of acidity and alkalinity, pH, at its unusual level of around 12 or 13 as the carbonation reaction occurs. However, eventually all the locally available calcium hydroxide reacts, precipitating the calcium carbonate and allowing the pH to fall to a level where steel will corrode. This is illustrated in Figure 7.5.

Carbonation damage occurs most rapidly when there is little concrete cover over the reinforcing steel. Carbonation can occur even when the concrete cover depth to the reinforcing steel is high.

This may be due to a very open pore structure where pores are well connected and allow rapid CO_2 ingress. It may also happen when alkaline reserves in pores are low. These problems occur when there is a low cement content, high water cement ratio, and poor curing of the concrete.

A carbonation front proceeds into the concrete roughly following the laws of diffusion. These are most easily defined by the statement that the rate is inversely proportional to the thickness:

$$Dx/dt = D_o/x \qquad (7.5)$$

where x is distance from concrete surface faced to environment, t is time, and D_o is the diffusion constant. The diffusion constant D_o is determined by the concrete quality. At the carbonation front, there is a sharp drop in alkalinity from pH 11 to 13 down to less than pH 8.

At the level, the passive layer, which was created by the alkalinity, is no longer sustained so corrosion proceeds by the general corrosion mechanism as described.

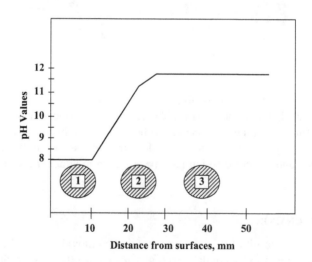

FIGURE 7.5 Relation between carbonation depth and level of pH values.

Many factors influence the ability of reinforced concrete to resist carbonation induced corrosion. The carbonation rate, or rather the time to carbonation induced corrosion, is a function of cover thickness, so good cover is essential to resist carbonation. As the process is none of neutralizing the alkalinity of the concrete, good reserves of alkali are needed, i.e., a high cement content.

The diffusion process is made easier if the concrete has an open pore structure. On the macroscopic scale, this means that there should be good compaction. On a microscopic scale, well cured concrete has small pores and lower connectivity of pores so the CO_2 has a harder job moving through the concrete.

7.3.2.1 Carbonation Transport Through Concrete

Carbon dioxide diffused through the concrete and the rate of movement of the carbonation front approximates to Fick's law of diffusion (Schiessl, 1988). This states that the rate of movement is inversely proportional to the distance from the surface as in Equation 7.6. However, as the carbonation process modifies the concrete pore structure as it proceeds, this is only an approximation. Crack changes in concrete composition and moisture levels with depth will also lead to deviation from the perfect diffusion equation. Integration of Equation (7.6) gives a square root law that can be used to estimate the movement of the carbonation front.

Empirically, a number of equations have been used to link carbonation rates, concrete quality, and environment. Table 7.7 summarizes some of those equations and shows the factor that has been included. Generally, there is a time dependence. As discussed above, the other factors are exposure, water/cement ratio, strength, and CaO content (functions of cement type and its alkali content).

For example, we consider the basic equations as follows:

$$d = A \cdot t^n \tag{7.6}$$

where d is the carbonation depth in millimeters, a is the a coefficient, t is the time (years), and n is an exponent, usually = 1/2.

A number of empirical calculations have been used to derive values of A and n based on such variables as exposure conditions (indoor and outdoor, sheltered or unsheltered), 28-day strength, and water/cement ratio as shown in Table 7.7.

Schiessel (1988) shows the relation between time and the depth of corrosion at different environmental conditions in Figure 7.6.

7.3.2.2 Parrott's Determination of Carbonation Rates from Permeability

For a new concrete mix or structure, the prediction of carbonation rate is complicated by the lack of data to extrapolate. In a series of papers (Parrott and Hong, 1991; Parrot, 1994a, b), a methodology was outlined for calculating the carbonation rate from air permeability measurements with a specific apparatus. Parrott (1987) analyzed the literature and suggested that the carbonation depth D at time t is given by

$$D = aK^{0.4}t^n - C^{0.5} \tag{7.7}$$

TABLE 7.7
A Selection of Carbonation Depth Equations (from Parrott, 1987)

Equation	Parameters
$d = A \cdot t^n$	d = carbonation depth
	t = time in years
	A = diffusion coefficient
	n = exponent (approximately ½)
$d = ABC \cdot t^{0.5}$	A = 1.0 for external exposure
	B = 0.07–1.0 depending on surface finish
	$C = R(wc - 0.25)/\left(0.3(1.15 + 3wc)\right)^{1/2}$
	R = coefficient of neutralization, a Function of mix design and additives
$d = A(Bwc - c)t^{0.5}$	A is a function of curing
	B and C are a function of fly ash used
$d = 0.43(wc - 0.4)\left(12(t-1)\right)^{0.5} + 0.1$	28 day cured
$d = 0.53(wc - 0.3)(12t)^{0.5} + 0.2$	uncured
$d = \left(2.6(wc - 0.3)^2 + 0.16\right)t^{0.5}$	sheltered
$d = (wc - 0.3)^2 + 0.07)t^{0.5}$	unsheltered
$d = 10.3\, e^{-0.123f28}$ at 3 years	unsheltered
	fX = strength at day X
$d = 3.4\, e^{-0.34f28}$ at 3 years	sheltered
$d = 680(f28 + 25)^{-1.5} - 0.6$ at 2 years	
$d = A + B/f28^{0.5} + c/(CaO - 46)^{0.5}$	CaO is alkali content expressed as CaO
$d = \left(0.508/f35^{0.5} - 0.047\right)(365t)^{0.5}$	
$d = 0.846\left(10wc/(10f7)^{0.5} - 0.193 - 0.076wc\right)(12t)^{0.5} - 0.95$	
$d = A\left((T - t_i)t^{0.75}\, C_1/C_2\right)^{0.5}$	t_i = induction time
	T = temperature in °K
	$C_1 = CO_2$ concentration
	$C_2 = CO_2$ bound by concrete

where K is air permeability (in units of $10^{-6}\,m^2$), c is the calcium oxide content in the hydrated cement matrix for the cover concrete, and a = 64. K can be calculated from the value at 60% relative humidity, r, by the equation:

$$K = m \cdot K60 \qquad (7.8)$$

where m = 1.6 – 0.0011 r – 0.0001475 r^2 or m = 1.0 if r < 60, and n is 0.5 for indoor exposure but decrease under wetted conditions to

$$n = 0.02536 + 0.01785\, r - 0.0001623\, r^2 \qquad (7.9)$$

Reliability of Concrete Structure

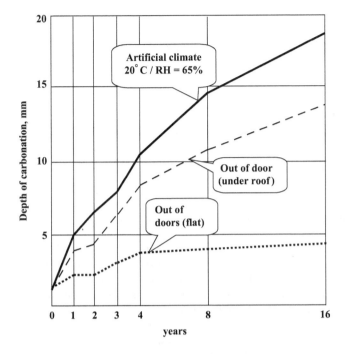

FIGURE 7.6 Effect of climatic conditions on the rate of carbonation.

Therefore, increasing the concrete cover depth is required to prevent carbonation from reaching the steel. The concrete cover can be calculated based on the measure of the air permeability and the relative humidity.

7.3.3 Corrosion Rates

For carbonation, it seems that the rate of corrosion falls rapidly as the relative humidity in the pores drops below 75% and rises rapidly to a relative humidity of 95% (Tutti, 1982).

For any sort of corrosion, there is also approximately a factor of 5–10 reduction in corrosion rate with 10°C reduction in temperature.

According to Schiessel (1988), the rate of corrosion in carbonated concrete is a function of relative humidity (RH), wetting or drying conditions, and the chloride content. Thus, the decisive parameters controlling corrosion in carbonated concrete will be associated with steady-state RH or wetting/drying cycles and conductivity increases associated with level of chloride in the concrete.

El Sayed et al. (1987) measured the corrosion rate for coated steel in tap water and in a solution of 1% sodium chloride (NaCl) plus 0.5% sodium sulfate (Na_2So_4). It can be seen that the coating decreases the value of the steel corrosion rate to about one tenth of the value obtained for the uncoated steel in both media. The corrosion rate for coated and uncoated steel bar is presented in Table 7.8.

TABLE 7.8
Corrosion Rates Measured by El-Sayed et al. (1987)

	Corrosion Rate (mm/year)	
Condition	Tape Water	1% NaCl + 0.5% Na$_2$So$_4$
Non-coated	0.0678	0.0980
Coated	0.0073	0.0130

However, the corrosion rate in most researches is within 0.015–0.09 mm/year (El Abiary et al. 1992).

7.3.4 STATISTICAL ANALYSIS OF INITIATION AND CORROSION RATES

In reliability analysis of concrete structures, many researches used different probabilistic models to describe initiation and corrosion rate of steel bars in concrete.

Mori and Ellingwood (1994a) used the Poisson process with parameters ν(t) to describe the initiation of corrosion following carbonation. The mean Poisson is the parameter ν(t) which is expressed as follows:

$$\nu(w) = \begin{cases} 0 & \text{for } w < t^* \\ \nu & \text{for } w \geq t^* \end{cases} \quad (7.10)$$

where t* is a deterministic time considered to be 10 years and ν is the mean initiation rate of corrosion considered equal to 0.2/year.

On the other hand, typical corrosion rates of steel in various environments have been reported in recent years.

According to Ting (1989), the average corrosion rate C_r for passive steel in concrete attacked by chlorides is about 100 µm/year.

According to Mori and Ellingwood (1994b), the typical corrosion rate, C_r, is a time-invariant random variable described by a lognormal distribution with mean C_r, of 50 µm/year, and coefficient of variation V_{cr} of 50%.

Because the corrosion rate changes with environment, no accurate data are available to predict the real corrosion rate.

Based on the average corrosion rates reported in Ting (1989), Mori and Ellingwood (1994b), and Frangopol et al. (1997), three corrosion rates of 64, 89, and 114 µm/year can be used to cover most cases of corrosion rates, which depends on different environment conditions.

These rates may suggest a mean (i.e., 89 µm/year) and a standard deviation (i.e., 25 µm/year).

7.3.5 CORROSION EFFECT ON SPALLING OF CONCRETE

Various efforts have been made to estimate the amount of corrosion that will cause spalling of concrete cover.

According to Broomfield (1997), the cracking is induced by less than 0.1 mm of steel bar section loss, but in some cases, far less than 0.1 mm has been needed. This is a function of the way that the oxide is distributed (i.e., How efficiency it stresses the concrete), the ability of the concrete to accommodate the stresses (By creep, Plastic or Elastic deformation), and the geometry of bar distribution that may encourage the crack propagation by concentrating stresses, as in the case of a closely spaced serious of bars near the surface, or at a corner where there is less confinement of the concrete to restrain cracking.

From the corrosion rate measurements, it would appear that about 10 µm section loss or 30 µm rust growth is sufficient to cause cracking.

However, rust is a complex mixture of oxides, hydroxides and hydrated oxides of iron having a volume ranging from twice to about six times that of the iron consumed to produce it.

According to El-Abiary et al. (1992), the time, t_s, in year, between initiation of corrosion and spalling of concrete is calculated from the following equations:

$$t_s = \frac{0.08 \cdot c}{D_b \cdot C_r} \qquad (7.11)$$

where c is the concrete cover in mm, D_b is the diameter of the steel bar, and C_r is the mean corrosion rate.

There was a study performed by Morinaga (1988) to investigate the conditions when concrete cover cracks due to corrosion of reinforcing steel; two series of experiments were carried out. One was to determine the tensile stress when concrete cover cracks. A hollow concrete cylinder was used as a specimen, simulating the internal diameter of the hollow cylinder as the diameter of the reinforcing steel and the wall thickness of the hollow cylinder as the thickness of cover concrete. Oil pressure was applied to the inner surface of the hollow, and the influence of bar diameter, cover thickness, and tensile strength of the concrete on the maximum oil pressure at failure of the specimen was investigated. The other was to determine the tensile strain when concrete cover cracks. A piece of reinforcing steel was embedded at the center of the concrete cylinder. The specimen was immersed in a salt solution and direct current was applied to the reinforcing steel and was forced to corrode electrolytically. After the concrete cylinder cracked, the reinforcing steel was weighed, and the amount of corrosion or the volume of corrosion products was determined.

Combining the results of these two experiments, the equation to estimate the amount of corrosion when concrete cover cracks due to corrosion was obtained as follows:

$$Q_{cr} = 0.602\left(1 + \frac{2c}{d}\right)^{0.85} d \qquad (7.12)$$

where
Q_{cr} is the amount of corrosion when concrete cracks ($\times 10^{-4}$ g/cm^2),
C is the cover thickness of concrete (mm), and
D is the diameter of reinforcing steel (mm).

7.3.6 CAPACITY LOSS IN REINFORCED CONCRETE COLUMNS

As shown in Chapter 4, the capacity of the reinforced concrete column is dependent on the cross-section dimensions (concrete and steel area) and material strength (concrete strength and steel yield strength).

In the case of uniform corrosion as shown in Figure 7.6, the total longitudinal reinforcement area can be expressed as a function of time t as follows:

$$A_s(t) = \begin{cases} n\pi D_b^2/4 & \text{for} \quad t \leq T_i \\ n\pi \left[D_b - 2C_r(t-T_i)\right]^2/4 & \text{for} \quad t > T_i \end{cases} \quad (7.13)$$

where D_b is the diameter of the bar, n is number of bars, T_i is time of corrosion initiation, and C_r is the rate of corrosion. Equation 7.7 takes into account the uniform corrosion propagation process from all sides (Figure 7.7).

7.4 PARAMETRIC STUDY FOR CONCRETE COLUMN

There are different parameters that affect the reliability of the reinforced concrete columns under corrosion attack. However, the corrosion itself depends on some variables related to environment and weather conditions.

Two types of reinforced concrete column sections' dimensions are taken into consideration. For the first one, the dimensions of length and width are equal to 50 × 50cm with longitudinal steel ratio equal to 1% where the second section is designed to carry the same load with section dimensions 40 × 40cm with longitudinal steel ratio equal to 4%.

The effect of percentage of longitudinal steel bars on the reliability is discussed. Two values for the steel ($\rho = 1\%$ and $\rho = 4\%$) are considered in the analysis to represent the range of allowable steel ratios in Egyptian code () for interior columns.

The environment is different from location site to other as dry and wetted conditions, which have impact on the corrosion rate. Therefore, three different corrosion rates (0.064, 0.089, 0.114 mm/year) are taken into consideration in the analysis, where these rates cover the most types of environment conditions (Frangopol et al., 1997).

The steel bars before using may have rust; on the other hand, the steel bars may take from manufacture to pouring the concrete with no adequate time to have a rust.

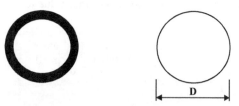

Uniform corrosion on the steel bars   Steel bars without corrosion

FIGURE 7.7 Reduction in steel diameter due to uniform corrosion.

Reliability of Concrete Structure

Therefore, in these two cases, the initial time of corrosion is different so different initial times (0, 10, 20 years) of corrosion are taken to study their effect on the reliability of columns.

The effect of age on concrete strength and correspondingly on reliability index is discussed in Section 7.4.1.

In Section 7.4.2, the effect of steel bar ratio on the reliability is discussed. The effect of corrosion rates values on reliability of column is presented in Section 7.4.3.

7.4.1 Effect of Age

The strength of concrete increases with age depending on the environmental condition as discussed in Section 7.2.

The reliability index is calculated in the case of dry condition and wet condition for $\rho = 1\%$ and $\rho = 4\%$ as shown in Figure 7.8. From this figure, one can find that, in general, the reliability index in the case of $\rho = 4\%$ is higher than that in the case of $\rho = 1\%$ for the first 10 years at wet and dry conditions. However, the reliability index after 10 years in the case of $\rho = 1\%$ with wet condition will be less than that for $\rho = 4\%$ at dry condition with about 4%.

From this figure, one can observe that in the case of dry condition the reliability index increases up to 1 year and after that it remains constant. Moreover, in the case of $\rho = 1\%$ the reliability index in wet condition increases after 2 years for that in dry condition; on the other hand, for $\rho = 4\%$ the increase in the value of reliability index for wet than that for dry condition will be after 4 years.

Therefore, the weather condition will be faster effect for case of $\rho = 1\%$ than that for case of $\rho = 4\%$, but after 10 years the percentage of increase of reliability index for wet to that for dry condition will be the same for the 2% of longitudinal steel bars.

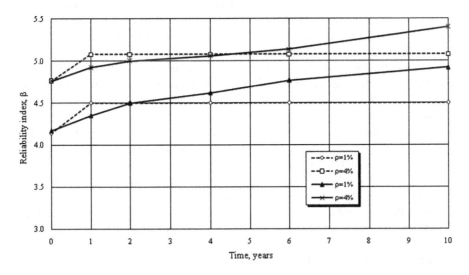

FIGURE 7.8 Effect of environmental conditions and steel reinforcement Ratios on the Reliability of reinforced concrete columns ignoring corrosion effect.

7.4.2 EFFECT OF PERCENTAGE OF LONGITUDINAL STEEL

According to clause 10.9.1 of the ACI code, the total area of the longitudinal reinforcement should not be less than 1% or more than 8% of the gross-sectional area. More comprehensive information in these regards is provided in BS 8115. According to Clause 3.12.6.2 of that code, the area of longitudinal reinforcement should not exceed the following amounts referred to area gross:

6% for vertically cast columns,
8% for horizontal cast columns,
10% at laps in vertically or horizontally cast columns.

An upper limit is imposed on the ratio of the longitudinal reinforcement to avoid undesirable congestion of bars and hence unsatisfactory compaction of concrete. A lower limit is imposed on the ratio of the longitudinal reinforcement to resist unavoidable bending moments, to curb the effects of creep and shrinkage of concrete, and to reduce the dimensions of the member because a part of the load will be carried by the steel with its much greater strength.

In this section, the effect of steel corrosion and the concrete strength along the lifetime of the structure on the reliability index of column with different steel ratios is discussed.

The corrosion rates 0.064, 0.089, and 0.114 mm/year are taken into consideration. These rates may suggest a mean (i.e., 0.089 mm/year) and a standard deviation (0.025 mm/year) (Frangopol et al. (1997).

Also, 2% of longitudinal steel bar in concrete is taken 1% and 4% which represents a wide range of percentage of longitudinal steel bars which are used in column design in Egyptian code and British standard.

The initial time of corrosion is assumed to be 10 years (Mori, 1994a, b).

After formulating the limit state equation as discussed in Chapter 4, the reliability index of column is calculated assuming three corrosion rates and percentage of steel bars at different lifetime using 10,000 trials of Monte-Carlo simulation.

The effect of the 2% of the steel bars on the reliability index at corrosion rate equal to 0.064 mm/year along the lifetime is shown in Figure 7.9. From this figure, one can find that the reliability index of the reinforced concrete column is decreased gradually after 10 years which is the time of starting corrosion.

Moreover, one can observe that the reliability index in the case of $\rho = 4\%$ is higher than that in the case of $\rho = 1\%$ up to 40 years, but after that the reliability index in the case of $\rho = 1\%$ is slightly higher than that in the case of $\rho = 4\%$.

It is also noticed that the reliability index of reinforced concrete column in the case of $\rho = 4\%$ is higher than that in the case of $\rho = 1\%$.

According to MacGregor (1976), a value of $\beta = 4.0$ is used in structural members when consequences of failure become a severe or the failure occurs in brittle manner.

Considering the value of $\beta = 4.0$ (suggested by MacGregor), one can find from Figure 7.9 that repair must be done for columns after 17 years in the case of $\rho = 1\%$, and after 30 year in the case of $\rho = 4\%$.

Reliability of Concrete Structure   165

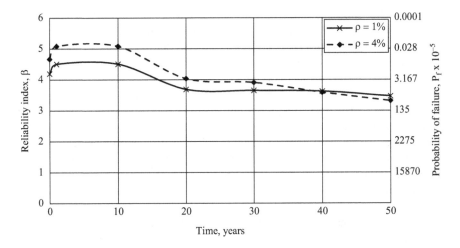

FIGURE 7.9 Effect of steel percentage on the reliability index of a reinforced concrete column at corrosion rate $C_r = 0.064$ mm/year.

From Figure 7.10, one can find that in the case of $\rho = 4\%$ the reliability index is higher than that at $\rho = 1\%$ until 37 years but after that the reliability index in case of $\rho = 1\%$ is higher than that in case of $\rho = 4\%$.

From this figure and by considering $\beta = 4.0$, one can find that the repair must be done after slightly less than 17 years in the case of $\rho = 1\%$ and at 20 years in the case of $\rho = 4\%$.

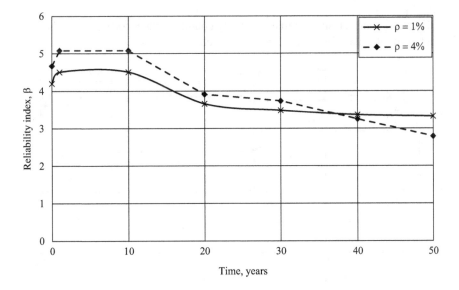

FIGURE 7.10 Effect of steel percentage on the reliability index of a reinforced concrete column at corrosion rate $C_r = 0.089$ mm/year.

In the case of corrosion rate 0.114 mm/year as shown in Figure 7.11, the reliability index in case of 4% is higher than that at $\rho = 1\%$ until around 30 years.

Considering $\beta = 4.0$ in Figure 7.11, one can find that the repair of the columns must be done after 15 years in the case of $\rho = 1\%$ and at 20 year in the case of $\rho = 4\%$.

Moreover, one can find that at the begging the probability of failure of the column with $\rho = 4\%$ is lower than that in the case of $\rho = 1\%$ until 30, 38, and 42 years with corrosion rates 0.064, 0.089, and 0.114 mm/year, respectively. Then the probability of failure will be higher in case of $\rho = 4\%$.

Therefore, the lower steel ratio recommended for the environmental condition causes high corrosion rate as the most column strength capacity is carried by concrete.

It is very important to discuss, also, the case when the steel bars at construction begin to corrode. Figure 7.12 illustrates this case in 2% of longitudinal reinforcement 1% and 4%.

It can be shown that in the case of $\rho = 4\%$ the reliability index is higher than that in the case of $\rho = 1\%$ until 20 years. In the case of $\rho = 4\%$ the rate of decreasing reliability is high.

7.4.3 Effects of Corrosion Rate

Reliability index is plotted in Figures 7.9 and 7.10 for different corrosion rates versus column ages.

These figures show that the reliability index for reinforced concrete columns increases as the corrosion rate decreases, for the two cases of steel ratios.

The three types of corrosion rates taken into consideration cover the most cases of environment conditions. It is also noticed that the reliability index of columns for later ages is highly affected by corrosion rate in the case of $\rho = 4\%$ than that in the case of $\rho = 1\%$.

Therefore, one can suggest that in wetting condition when corrosion rate is high it is better to take a minimum percentage of steel.

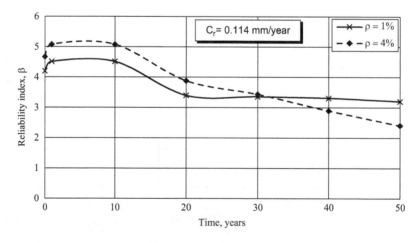

FIGURE 7.11 Effect of steel percentage on the reliability index of a reinforced concrete column at corrosion rate $C_r = 0.114$ mm/year.

Reliability of Concrete Structure

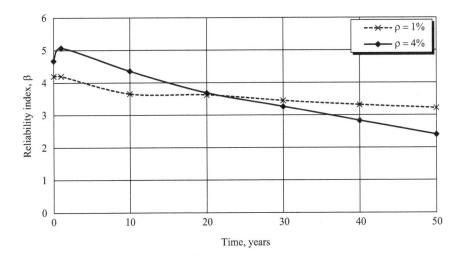

FIGURE 7.12 Effect of percentage of steel on reliability index at corrosion initiation time zero.

7.4.4 Effect of Initial Time of Corrosion

Initial time of corrosion is the time from constructing the column until the carbonation of the cover reaches the reinforcement and then the pH will be less than 5 in the longitudinal reinforcement. Then the corrosion will propagate inside the steel bar.

In this section, three values of initial time of corrosion T_i are considered (Figures 7.13 and 7.14):

$T_i = 0$ indicates that the outer surfaces of the bars have been already begun to corrode at construction and

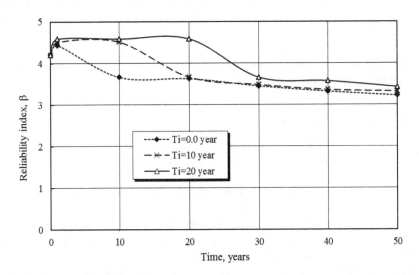

FIGURE 7.13 Effect corrosion initial time on the reliability index at reinforcement ratio $\rho = 1\%$.

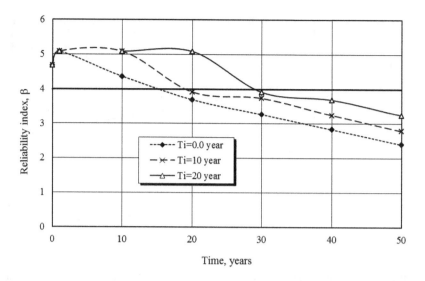

FIGURE 7.14 Effect corrosion initial time on the reliability index at reinforcement ratio $\rho = 4\%$.

$T_i = 10$ years indicates that corrosion starts after 10 years which represent the normal condition, while $T_i = 20$ years indicates that corrosion starts after 20 years.

The third case ($T_i = 20$ years) represents a thick concrete cover which takes time to reach pH less than 5. The time of beginning of corrosion depends on the thickness of concrete cover, the grade of concrete, and proportional to that density of concrete.

From Figures 7.13 and 7.14, one can find that the reliability of reinforced concrete column increases by increasing the initial time of corrosion. However, the reliability index values are slightly different in the three initial times after 40 years in the case of $\rho = 1\%$ but in the case of $\rho = 4\%$ there is a high reliability range between initial time 20 years and initial time 0.

Therefore, the parameter which increases the initial time is very important to take into consideration in the case of $\rho = 4\%$.

7.4.5 Effect of Eccentricity

The value of eccentricity (e = 0.05 h) is added to the simulation program to study the effect of different parameters on the reliability of reinforced concrete column with eccentric load.

The effect of eccentricity on the reliability index in the case of corroded bars with time is presented in Figures 7.15 and 7.16.

In these figures, the reliability indices with time are represented in the case of $\rho = 1\%$ and $\rho = 4\%$, respectively, with two values of extreme of eccentricity specify by the code Egyptian code and comply with BS.

In these two figures, one can see that, in general, the reliability index in the case of no eccentricity is higher than that in the case of eccentricity (e = 0.05 h).

Reliability of Concrete Structure

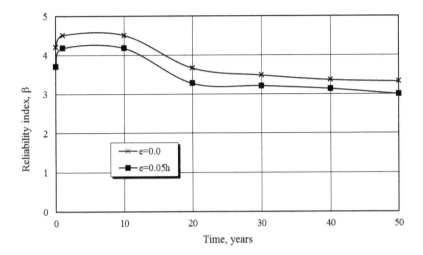

FIGURE 7.15 Effect of eccentricity on the reliability index with time ($\rho = 1\%$).

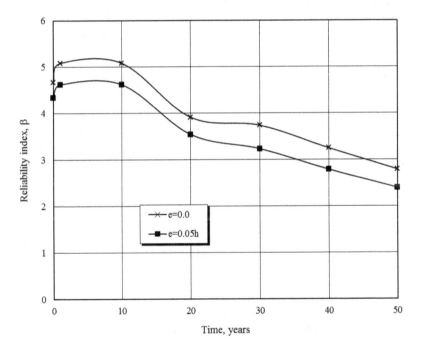

FIGURE 7.16 Effect of eccentricity on the reliability index with time ($\rho = 4\%$).

Figures 7.17–7.20 show the effect of reliability index with time in the case of eccentrically loaded column (e = 0.05 h) for different corrosion rates.

From these figures, one can see that repair begins when the reliability index reaches to 4.0 (specify by MacGregor). In the case of corrosion rate 0.064 mm/year, the repair is better to begin at 13 years for $\rho = 1\%$ and 16 years for $\rho = 4\%$.

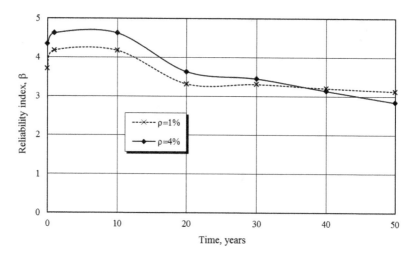

FIGURE 7.17 Effect of reinforcement ratio on the reliability index in case of small eccentricity (e = 0.05 h) and corrosion rate C_r = 0.064 mm/year.

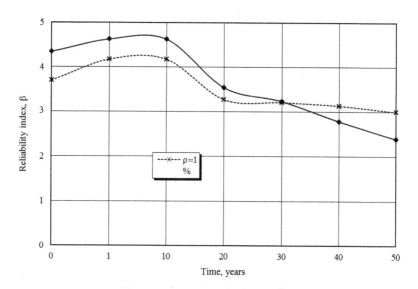

FIGURE 7.18 Effect of reinforcement ratio on the reliability index in case of small eccentricity (e = 0.05 h) and corrosion rate C_r = 0.089 mm/year.

At the rate of corrosion equal to 0.089 mm/year, the repair time should be slightly less than the previous rate of corrosion in the two values of steel percentages.

However, at the rate of corrosion equal to 0.114 mm/year the repair needs to be at 12 years for ρ = 1% but at 15 years for ρ = 4%.

From the previous results, one can conclude that repair of column with e = 0.05 h should be done at least after 12 years and at most 16 years according to the corrosion rate and steel ratio.

Reliability of Concrete Structure

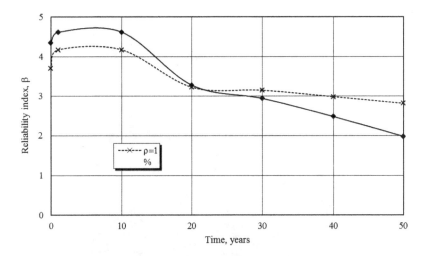

FIGURE 7.19 Effect of reinforcement ratio on the reliability index in case of small eccentricity (e = 0.05 h) and corrosion rate C_r = 0.114 mm/year.

FIGURE 7.20 Corrosion on the main girder and columns for a bridge.

On the other hand, the repair of concentrically loaded column should be done at least after 15 years and at most 37 years according to the corrosion rate and steel ratios.

7.5 EFFECT OF CORROSION ON THE GIRDER

The deterioration of RC structures due to reinforcement corrosion has been a major problem in the past 20 years. According to Kilareski (1980), most of the deterioration of the reinforcing-steel corrosion is occurred due to chloride ion contamination of the concrete.

Figure 7.20 illustrates a corrosion of the steel bars for the main girder and the column of a bridge. Figure 7.21 presents a collapse of the building due to severe corrosion on the columns due to poor maintenance.

FIGURE 7.21 Building collapse in Egypt in 1993.

Kilareski (1980) describes this corrosion process as follows:

Under normal environmental conditions, steel-reinforcing bars embedded in concrete do not corrode. Usually a thin film of iron oxide is present on the surface of the rebar when it is encased in the concrete. The high pH environment (approximately 13) associated with the hydration of the Portland cement is usually sufficient to keep the protective film stable. However, sufficient concentrations of chloride ions can lower the pH; and if moisture and oxygen are present, the rebar can begin to corrode. The chloride ions are provided by the de-icing salts used on the highway system in the winter months. After a few winter seasons, there are usually enough chloride ions at the level of the rebar to break down the passive environment around the steel. Once the rebar begins to corrode and build up the red rust by-product of corrosion, it is only a matter of time before enough force is generated so that a spall or pot hole is formed.

In this study, corrosion is seen as a two-phase process. The first phase spans from the time of construction to the time of corrosion initiation, and the second phase follows until unacceptable levels of section loss have occurred. Chloride ions are the most common degradation agents. A high level of chloride concentration leads to a breakdown of the protective passivation layer surrounding the steel reinforcement.

According to West and Hime (1985), corrosion initiation occurs with chloride concentrations (at the level of the rebar) of about 0.83 kg/cu m (1.4 lb/cu yd), for typical mixes of normal weight concrete of about 2,300 kg/cu m (145 lb/cu ft). The presence of large chloride concentrations at the surface of the concrete will cause the chloride concentrations at the rebar level to increase over time. The corrosion does not occur until the accumulation of chloride at the rebar surface exceeds the threshold value.

Reliability of Concrete Structure

Studies indicate that chloride penetration (Cady and Weyers, 1984; Takewaka and Matsumoto, 1988) can be treated as a diffusion process and seems to follow Fick's law of diffusion

$$\frac{\partial C_x}{\partial t} = \frac{D \partial^2 C_x}{\partial x^2} \tag{7.14}$$

where C_x = chloride ion concentration at distance x from the concrete surface after time t of exposure to chloride sources;

D is the effective chloride diffusion coefficient of concrete and t is time. Depending on the boundary conditions and amount of chloride, several possible solutions to this differential equation are possible.

It was found (Takewaka and Matsumoto, 1988) that the effective chloride diffusion coefficient of concrete D ($in^2./s$) depends on the water-to-cement ratio, w/c, and the type of cement as follows:

$$D = D_{w/c} \cdot D_c \tag{7.15}$$

In Equation 7.15, the coefficient denoting the effect of water-to-cement ratio, $D_{w/c}$ (sq in./s), is obtained from the following equation:

$$D_{w/c} = 10^{-6.274 - 0.076 w/c + 0.00113(w/c)^2} \tag{7.16}$$

where w/c is in percent and the coefficient denoting the effect of the cement D_c is 1.2, 1.0, 0.3, and 0.08 for high early strength Portland cement, ordinary Portland cement, blast furnace slag cement, and alumina cement, respectively.

During the winter season, the chloride ions are supplied by the deicing salts used on highway systems. After the winter season, there are usually enough chloride ions at the level of the rebar to induce corrosion. The total deicing salts commonly used in the winter season are considered applied once a year. For this case, the following equation will be used:

$$C_{x,t} = \frac{G}{\sqrt{\pi D_c t}} \exp\left(\frac{-x^2}{4 D_c t}\right) \tag{7.17}$$

where C_x is the chloride concentration at depth x at time t and G is the surface chloride content representing the amount of chloride deposited on the concrete surface (lb/cu yd).

In the case of the severe marine structure according to Gjorv (2009), the mean and standard deviation for the chloride concentration is presented in Table 7.9.

Table 7.9 is used as a general guide for the chloride content mean value and standard deviation with different chloride effect level that will be useful to apply in the reliability of the reinforced concrete structure under chloride attack.

To evaluate the probability of the corrosion occurring depends on the percentage of chloride content by weight of cement or concrete as shown in Figure 7.10. So from this table we can evaluate and define the concrete structure service life against chloride attack (Table 7.10).

TABLE 7.9
General Guide for Estimation of Chloride on Concrete Structures in Severe Marine Environments

	Cs (% by wt. of Cement)	
Chloride Load	Mean Value	Standard Deviation
High	5.5	1.3
Average	3.5	0.8
Moderate	1.5	0.5

TABLE 7.10
Risk for Corrosion Depending on the Chloride Content

Chloride Content (%)

By Weight of Cement	By Weight of Concrete	Risk of Corrosion Category
>2.0	>0.36	Certain
1.0–2.0	0.18–0.36	Probable
0.4–1.0	0.07–0.18	possible
<0.4	<0.07	Negligible

The concrete structure is usually exposed to chloride, as shown in Figure 7.22, presents the corrosion of the steel bars and falling of concrete cover for concrete structure in a jetty for a harbor.

On the other hand, the concrete cover is a major influence on the corrosion which is clearly obvious in Figure 7.23 which presents the relation between the nominal concrete cover thickness and the probability of corrosion occurring. It can be found that after 40 years in case of concrete cover equal to 120 mm the probability of corrosion occurring is approximately zero; however, in the case of concrete cover thickness

FIGURE 7.22 Corrosion of a jetty.

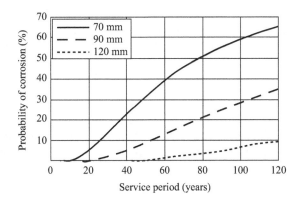

FIGURE 7.23 Effect of increased nominal concrete cover on the probability of corrosion occurs.

equal to 90 mm the probability will increase to be 5%. If there is a reduction on concrete cover to be 70 mm, the probability can be increased to be around 22%.

Equation 7.17 is best suited for concrete structures exposed to a single application of chloride. Using the method of superposition is well suited for concrete structures that have been exposed to repeated applications of chloride deposits such as highway RC bridges.

As discussed by Frangopol et al. (1997), in the case of bridge girder the change in chloride concentration in both space (i.e., penetration in concrete) and the values considered for G and D in Equation 7.17 are 1.485 kg/m^3 (2.5 lb/cu yd) and 3.26×10^{-8} cm^2/s (5.06×10^{-9} in^2/s), respectively. It is found that after a single application (at time t = 0) the chloride concentration decreases with time near the concrete surface, penetration below 1.8 cm (0.7 in.). By comparing the actual chloride concentration at the surface of reinforcement with the threshold concentration for initiation of corrosion approximately, 0.83 kg/cu m (1.4 lb/cu yd), it is easy to find the corrosion initiation time required to depassivate the surface of the reinforcement. It should be mentioned that it may be modified to account for other factors that can influence the corrosion initiation time, such as repeated wetting and drying of the concrete at the surface. However, reliable data on the effects of these factors are not available.

Typical corrosion rates of steel in various environments have been reported in recent years. According to Ting (1989), the average corrosion rate v for passive steel in concrete attacked by chlorides is about 100 μm/year (0.004 in/year). From Mori and Ellingwood's research (1994b), the typical corrosion rate is a time-invariant random variable described by a lognormal distribution with mean, 50 μm/year (0.002 in./year) and coefficient of variation of 50%. Because the corrosion rate changes with the environment, no accurate data is available to predict the real corrosion rate. Based on the average corrosion rates reported in Ting (1989) and Mori and Ellingwood (1994b), three corrosion rates of 64, 89, and 114 μm/year (0.0025, 0.0035, 0.0045 in./year) are used in this study to cover most cases. These rates may suggest a mean (i.e., 89 μm/year) and a standard deviation (i.e., 25 μm/year).

As corrosion gradually progresses, the remaining capacities in both bending and shear decrease. The flexural strength of an isolated reinforced concrete beam at time t, Mr(t), may be computed as the fundamental of concrete beam design as in Figure 7.24:

$$M_r(t) = A_s(t)f_y\left(d - \frac{a}{2}\right) \quad (7.18)$$

where f_y is the yield strength of a reinforcing bar, d is the effective depth, and a is the depth of the equivalent rectangular stress block. On the other hand, the time-dependent shear strength is

$$V(t) = V_c + V_s(t) \quad (7.19)$$

In the above equation, it is assumed that the shear strength of concrete V_c is time-independent. When only shear reinforcement perpendicular to the axis of the member is used, the shear strength due to stirrups is given as follows:

$$V_s(t) = \frac{A_s(t)f_y d}{s} \quad (7.20)$$

where $A_v(t)$ is the area of the shear reinforcement within the distance and s is the spacing between stirrups.

The loss of material due to corrosion causes a reduction in both bending and shear capacities of RC beams. Based on 7.18 and 7.19, and AASHTO (Standard 1992) requirements for the design of RC T-bridge girders, the moment and shear capacities can be calculated for both time-independent (intact) and time-dependent (damaged under corrosion) limit states.

The average bridge in USA is now 43 years old. According to the U.S. Department of Transportation, of the 600,905 bridges across the country as of December, 2008, 72,868 (12.1%) were categorized as structurally deficient and 89,024 (14.8%) were categorized as functionally obsolete. From 2005 to 2008, the number of deficient (structurally deficient plus functionally obsolete) bridges in rural areas declined by 8,596. However, in urban areas during the same time frame, there was an increase of 2,817 deficient bridges. Put another way, in 2008 approximately one in four rural

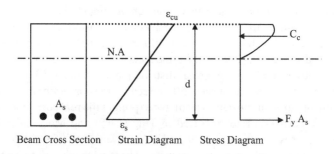

FIGURE 7.24 Strain and stress diagrams of the beam cross-section.

Reliability of Concrete Structure

bridges were deficient, while one in three urban bridges were deficient. The urban impact is quite significant given the higher level of passenger and freight traffic.

A structurally deficient bridge may be closed or restrict traffic in accordance with weight limits because of limited structural capacity. These bridges are not unsafe, but must post limits for speed and weight. A functionally obsolete bridge has older design features and geometrics, and though not unsafe, cannot accommodate current traffic volumes, vehicle sizes, and weights. These restrictions not only contribute to traffic congestion but also cause such major inconveniences as forcing emergency vehicles to take lengthy detours and lengthening the routes of school buses.

With truck miles nearly doubling over the past 20 years and many trucks carrying heavier loads, the spike in traffic is a significant factor in the deterioration of America's bridges. Of the more than 3 trillion vehicle miles of travel over bridges each year, 223 billion miles due to trucks moving.

To address bridge needs, states use federal as well as state and local funds. According to the American Association of State Highway and Transportation Officials (AASHTO), a total of $10.5 billion was spent on bridge improvements by all levels of government in 2004. Nearly half, or $5.1 billion, was funded by the Federal Highway Bridge Program—$3.9 billion from state and local budgets and an additional $1.5 billion in other federal highway aid. AASHTO estimated in 2008 that it would cost roughly $140 billion to repair every deficient bridge in the country about $48 billion to repair structurally deficient bridges and $91 billion to improve functionally obsolete bridges.[1]

Simply maintaining the current overall level of bridge conditions, i.e., not allowing the backlog of deficient bridges to grow, would require a combined investment from the public and private sectors of $650 billion over 50 years, according to AASHTO, for an average annual investment level of $13 billion. The cost of eliminating all existing bridge deficiencies as they arise over the next 50 years is estimated at $850 billion in 2006 dollars, equating to an average annual investment of $17 billion.[3]

7.6 RECOMMENDATIONS FOR DURABLE DESIGN

The reliability of reinforced concrete columns under corrosion attack is discussed taking into consideration 2% of longitudinal steel bars ($\rho = 1\%$ and $\rho = 4\%$), different corrosion rates, different rates of carbonation which related to different corrosion initial time, and the increase of concrete strength with time. Moreover, the effect of corrosion on the reliability index is discussed.

From the study that was performed by El-Reedy et al. (2000), one can conclude the following:

- The case of $\rho = 4\%$ has a higher reliability index than that for the case of $\rho = 1\%$.
- The case of wet condition has a higher reliability index than that for the case of dry condition.
- The reliability index will increase about 10% in the case of dry condition after 1 year.

- The repair of concentrically loaded column should be done at least after 15 years and at most 30 years according to the corrosion rate and steel ratio.
- The reliability index increases by increasing the initial time of corrosion and the initial time has more serious effect in case of $\rho = 4\%$.
- The repair of column with $e = 0.05\,h$ should be done at least after 12 years and at most 16 years according to the corrosion rate and steel ratio.

BIBLIOGRAPHY

AASHTO, 1992, *Standard Specifications for Highway Bridges*, 15th Edition, American Association of State Highway and Transportation Officials (AASHTO), Washington, D.C.

American Association of State Highway and Transportation Officials (AASHTO). Bridging the Gap, 2008.

Baykof, F., and Sigalof, Y., 1984, *Reinforced Concrete Structure*, Mier, Mosco.

Broomfield, J.P., 1997, *Corrosion of Steel in Concrete*, E&FN Spon, London.

Cady, P.D., and Weyers, R.E., 1984, Deterioration rates of concrete bridge decks, *Journal of Transportation Engineering*, Vol. 110, No. 1, pp. 34–44.

El-Reedy, M.A., Ahmed, M.A., and Khalil, B.A., Reliability analysis of reinforced concrete column under corrosion attack, *Proceedings of ACI 5th International Conference*, Cancun, Dec, 2002, pp. 77–91.

El-Sayed, H.A., Kamal, M.M., El-Ebiary, S.N., and Shahin, H., 1987, Effects of reinforcement corrosion and protective coatings on the strength of the concrete/steel bond, *Corrosion Prevention and Control*, Vol. 34, No. 1, pp. 18–23.

Frangopol, D.M., Lin, K.Y., and Allen, C., 1997, Reliability of reinforced concrete girders under corrosion attack, *Journal of Structural Engineering*, Vol. 123, No. 3, pp. 286–297.

Kilareski, W.P., 1980, Corrosion induced deterioration of reinforced concrete-An overview, *Materials Performance*, Vol. 19, No. 3, pp. 48–50.

MacGregor, J.G., 1983, Load and resistance factors for concrete design, *ACI Journal*, Vol. 80, No. 4, pp. 279–287.

Mori, Y., and Ellingwood, B.R., 1994a, Maintaining reliability of concrete structures I: Role of inspection/repair, *Journal of Structural Engineering*, Vol. 120, No. 3, pp. 824–845.

Mori, Y., and Ellingwood, B.R., 1994b, Maintaining reliability of concrete structures II: Role of inspection/repair. Journal of Structural Engineering, Vol. 120, No. 3, pp. 846–862.

Morinaga, S., Prediction of service lives of reinforced concrete buildings based on rate of corrosion of reinforcing steel, special report of institute of technology, Shimizu Corporation, No.23, Tokyo, Japan, 1988.

Schiessl, P., Corrosion of steel in concrete, Report of the Technical Committee 60-CSC, Chapman and Hall, London, 1997.

Takewaka, K., and Matsumoto, S., 1980, Quality and cover thickness of concrete based on the estimation of chloride penetration in marine environments, In: *Concrete in marine Environment*, SP-109, V.M. Malhotra, (ed.), American Concrete Institute (ACI), Detroit, MI, pp. 381–400.

Ting, S.C., The effects of corrosion on the reliability of concrete bridge girder, Ph.D. thesis, Department of Civil Engineering, University of Michigan, Ann arbor, MI, 1989.

Washa, G.W., Saemann, J.C., and Cramer, S.M., 1989. Fifty-year properties of concrete made in 1937, *ACI Material Journal*, Vol. 86, No. 4, pp. 367–371.

West, R.E., and Hime, W.G., 1985, Chloride profiles in salty concrete, Materials Performance, Vol. 24, No. 7, pp. 29–36.

8 Inspection Methodology
Visual Inspection

8.1 INTRODUCTION

Normally, the main cause of deterioration in the reinforced concrete structure in oil and gas facilities is due to corrosion. This appears obviously in buildings near to the seas and also in cold climates as they usually deice by using salt.

Before repairing the corrosion in concrete structure, the building must be evaluated with high accuracy, and the environmental conditions surrounding the building must also be identified.

After performing the risk assessment to be correct and technically accurate to determine the degree of building risk, the minor repair or rehabilitation process will be technically accurate and adequate, and more precisely.

On the other hand, if the diagnosis of disease is found to be correct, then the preferred medicine will also be correct. The preliminarily steps of a building assessment are the diagnosis of the causes of defects in the structure as a result of corrosion and the recognition of the causes leading to the corrosion. Subsequently, calculation of the service life is mainly dependent on a correct structure assessment.

The process of assessing the structures can be classified into two stages: the first one is the initial assessment of the building and is used to define the problem and develop the plan for a detailed evaluation of the structure to be carried out by visual inspection. The second step is to perform nondestructive testing and start evaluation calculation step.

One must carefully define the problem and the reasons while performing a detailed assessment of the building. At that stage, the detailed inspection of the whole building will be carried out. There has been a basis established for the inspection of the buildings for assessment in the Technical Report 26 (Concrete Society, 1984), as well as the American Concrete Institute.

The reasons for the deterioration of the reinforced concrete structures ranging from the presence of the cracks to the falling the concrete cover include: increasing stresses, plastic shrinkage, frost or cracked concrete for plastic concrete, move the wood form during construction, and the presence of an aggregate that has the capability to shrink or deteriorate as a result of the aggregate being contaminated with the alkalinity.

There are also some structures, such as reinforced concrete pipelines in sanitary projects and underground concrete structures like tunnel that come in direct contact with the soils. In this case, the sulphate present in the water will affect the concrete.

As we mentioned earlier, there are several types of cracks. Therefore, the engineer who is responsible for performing the evaluation must be highly qualified in that area so that he can specify precisely why the incidence of deterioration in concrete was observed with the incorrect assessment, which will be result in a wrong repair causing a heavy financial loss, and which will affect the structure safety.

The evaluation of structure is not only a structural assessment but also it is an assessment of the state of concrete in terms of the presence of corrosion and the rate of corrosion and collapse of concrete. As per previous collected data, we can decide if the concrete member can withstand loads.

8.2 CONCRETE STRUCTURE INSPECTION

This inspection process identifies clearly the geographical location of the structure, the nature and circumstances of weather conditions surrounding the building, the method of constructing the building its structural system, and the method of loading.

This assessment is often performed preliminary by visual inspection, concentrated on the cracks and the fall of concrete cover and also collecting data about the thickness of the concrete cover and the nature of structure in terms of the quality of construction and structure system whether it consists of beam and column system or slab on load bearing masonry or it is prestressed or precast concrete.

All such information is preliminary and it is necessary to perform some simple measurements, for example, to determine the depth of the chloride in the concrete. Take a sample from concrete that has collapsed and conduct a laboratory test.

The safety of the structure must be calculated precisely, especially after decreasing the cross-sectional area of steel due to corrosion, as well as the fall of the concrete cover, which reduces the total cross section area of the concrete member, as these two areas are the main coefficients that affect directly on the capacity of the concrete member to carry the loads.

8.2.1 COLLECTING DATA

The following data must be collected before performing the visual inspection and must be mentioned in the structure assessment report:

- Construction year.
- The data collected through contractor and engineering office.
- The structure system.
- The drawings and project specifications.
- The construction method.
- Concrete tests if available.
- The environmental condition (near sea or chemical factory).

All the above data must be provided to you by the owner if you are a third-party expert. For the new structure, these data are available. However, in most cases these data cannot be collected when evaluating an old structure that may be more than 20 or 30 years old. Thus, the drawings and specifications will not be available at all for a 20 year- old structure as there were no computer programs at that time.

Inspection Methodology

In most cases, all the hardcopy drawings will be depleted or will be in bad condition. However, if the owner has a big organization with an engineering department there may be a document control system and possibly you can find the drawings, but for the residential buildings you will be very lucky if the owner has these drawings.

The construction year is the very critical information as from this you can define the codes and specifications that were applied in that year and this depends on your experience.

Knowing the construction year, the contractor who constructed the building and the engineering office that provided the engineering design, one can imagine the condition of the building based on the repetition of the contractor and the engineering office, and thus one can identify the level of quality.

The construction system can be determined by visual inspection, through drawings, or by talking with the users of the building and the engineering office that reviewed the engineering design documents or supervised the construction.

The information about environmental conditions will be provided to you by the owner and you can find these conditions on the site and collect the data about the environmental condition from other sources, taking into consideration all the data, such as temperature in summer, winter, morning, night, and the wind and wave effect for offshore structure.

If you are an owner, you must be sure that all the data must be delivered to the consultants who perform the inspection, as any data are valuable. The engineering work is like a computer—good in, good out, and rubbish in, rubbish out. So you must verify all the data that you deliver to the expert who performs the building assessment.

The operation of the building and the main entrance management should be known and identified. For example, in some countries, the change of use from residential building to hospital or office building is applicable without getting approval from the governmental authority, whereas in the other countries it is necessary to have an approval before change of use and other clients special in industrial sector that follow ISO9001 should have a management of change policy.

If the data of a seismic zone, wind, and wave do not exist or are not known, you must obtain it from a third party who specializes in meeting the ocean criteria. This data are very important for offshore structures or marine structure whose assessment depends on these data if wind and wave represent the most loads affecting the structure.

The search for the original design package is very important. Most likely you will find the architectural, structural, mechanical, electrical, and plumbing construction drawings. In minor cases, especially for older buildings, you may likely find the structural design calculation.

Generally, in any building, the concrete compressive strength test results is performed by cylinder specimens in American code or cube specimens in the British standard, but the problem that one can face repeatedly is to find these data for structure after 20 years or more.

If these data are not available, the test must be informed by using an ultrasonic pulse velocity test, Schmidt hammer, or core test that will be described later.

8.2.2 Design Code

The reinforced concrete structures are generally designed in accordance with national or international consensus codes and standards, such as ACI 318, Eurocode 2, and Comite Euro International du Beton (1993). These consensus codes are developed and based on the findings of research and laboratory tests. Although present design procedures for concrete are dominated by analytical determinations based on strength principles, designs are increasingly being refined to address durability requirements (e.g., resistance to chloride ingress and improved freezing-and-thawing resistance). Inherent with design calculations and construction documents developed in conformance with these codes is a certain level of durability, such as requirements for concrete cover to protect embedded steel reinforcement under aggressive environmental conditions. Although the vast majority of reinforced concrete structures have met and continue to meet their functional and performance requirements, numerous examples can be cited where structures, such as pavements and bridges, have not exhibited the desired durability or service life. In addition to material selection and proportioning to meet concrete strength requirements, a conscious effort needs to be made as stated by Sommerville (1986) to design and detail pavements and bridges for long-term durability. A more holistic approach is necessary for designing concrete structures based on service-life considerations. Therefore, it is important to keep focus on the environmental and structural loading considerations, as well as their interaction, and design and construction influences on the service life of structures.

AASHTO (1991) specifies a 75-year design life for highway bridges. ACI 318 makes no specific life-span requirements. Other codes, such as Eurocode, are based on a design life of 50 years, but not all environmental exposures are considered. ACI 318 addresses serviceability through strength requirements and limitations on service load conditions. Examples of service load limitations include mid span deflections of flexural members, allowable crack widths, and maximum service level stresses in prestressed concrete. Other conditions affecting service life are applied to the concrete and the reinforcement material requirements and detailing. These include an upper limit on the concrete water cement ratio (w/c), a minimum entrained-air content depending upon exposure conditions, and concrete cover over the reinforcement.

Most international design codes and guidelines have undergone similar changes in the past 30 years. For example, concretes exposed to freezing and thawing in a moist condition or to deicing chemicals, ACI 318-63 allowed a maximum *w/c ratio* of 0.52 and air entrainment, while ACI 318-89 allows a maximum *w/c ratio* of 0.45 with air entrainment. In 1963, an appendix was added to ACI 318 permitting strength design. Then in 1971, strength design was moved into the body of ACI 318, and allowable stress design was placed into the appendix.

The use of strength design provided more safety and it was possibly more cost-effective to have designs with a known, uniform factor of safety against collapse, rather than designs with a uniform, known factor of safety against exceeding an allowable stress. Realizing that design by strength limits alone could lead to some unsuitable conditions under service loads, service-load limitations listed above were adopted in ACI 318. The service-load limitations are based on engineering

Inspection Methodology

experience only and not on any rigorous analysis of the effects of these limitations on the service life of the structure.

Load and resistance factor-strength design which is known as limit state design methods consider the loads applied to the structure and the resistance capacity of the structure to be two separate and independent conditions. The premise is that the strength of the structure should exceed the effects of the applied loads, which can be expressed as follows:

$$\text{Capacity} > \text{demand (over the desired service life).}$$

8.2.3 Visual Inspection

Visual inspection is the first step in any process of technical diagnosis and it starts by a general view for the structure as a whole and then concentrates on the general defects and after that precisely on the cracks and the deteriorated part and if it is due to the corrosion and the extent of corrosion on the steel reinforcement.

As we have stated earlier, the assessment of the building must be performed by an expert because the cracks in the concrete structure may not be the cause of corrosion as corrosion is not the only factor that causes cracks. However, it is the main factor for major deteriorated structures.

The purpose of the detailed inspection is to determine accurately whenever possible the degree of seriousness and the deterioration of the concrete.

Therefore, we need to know the amount of collapses that have occurred, the cause of the deterioration in the concrete, and the amount of repair that will be needed at this stage, because such quantities will be placed in the tender that will be put forward to contractors for minor or major repair.

At that stage, we will need to be familiar with the reasons for the deterioration in detail and have the capability to perform failure analysis technique.

Initially, visual inspection may be carried out in conjunction with the use of small hammer before carrying out other measurements to determine the depth of the carbon transformation in the concrete, as well as the degree of steel corrosion in concrete and how much it extends in the steel bars. Moreover, at this stage it is important to define the degree of concrete electric resistivity to predict the corrosion rate. All of these measurements will be discussed in this chapter.

As you should know, the weather condition that affects the building is the main factor that affects the measurement readings; these factors also affect the selection of the method of repair.

Determining the existing performance characteristics and extent and causes of any observed distress is accomplished through a condition assessment by personnel having broad knowledge in structural engineering, concrete materials, and construction practices.

The condition assessment commonly uses a field survey involving visual examination and application of nondestructive and destructive testing techniques if required, followed by laboratory and structure analysis studies.

Before conducting a condition assessment, a definitive plan should be developed to optimize the information obtained. The condition assessment begins with a review of the as-built drawings if they exist, or construction drawings and other information pertaining to the original design and construction, so this information, such as accessibility and the position of embedded-steel reinforcement and plates in the concrete, are known before the site visit. Next, a detailed visual examination of the structure is conducted to document information that could result from or lead to structural distress, such as cracking, spalling, leakage, and construction defects, such as honeycombing and cold joints, in the concrete. It is important to have photographs or video recordings made during the visual examination, which can provide a permanent record of this information.

After the visual survey has been completed, the need for additional surveys, such as delamination plane, corrosion, or pachometer, is determined. Results of these surveys are used to select portions of the structure to be studied in greater detail. Many of the investigation techniques with nondestructive section will be discussed in the later section. Any elements that appear to be structurally marginal, due to either unconservative design or effects of degradation, are identified and appropriate calculation checks made. After all the data and information have been collated and studied, a report is prepared.

Once the critical structural components have been identified through the condition assessment, a structural assessment can be required to determine the current condition, to form the basis for estimating future performance or service life, or both. As part of the assessment, it is important to note irregularities or inconsistencies in properties of materials, in design, in construction and maintenance practices, and the presence and effects of environmental factors. Although the assessment of a structure involves more than its load-carrying ability, an assessment of structural demand versus capacity is the first step. Performance requirements other than structural capacity are then addressed through supplementary tests to establish characteristics, such as leakage rate or permeability.

There is evidence of possible structural weakness (e.g., excessive cracking or spalling). The building or a portion of it has undergone general or local damage due to environmental or earthquake effect. There is doubt concerning the structure's capacity, and portions of a building are suspected to be deficient in design, detail, material, or construction.

An analytical assessment is recommended when sufficient background information is not available, such as sectional characteristics, material properties, and construction quality. In most cases, a static load test is impractical because of the test complexity or magnitude of the load required, sudden failure during a static load test can endanger the integrity of the member or the entire structure, or it is required by an authority.

The non-destructive test which will be described in Section 8.8 is very important to define the concrete strength and the condition of the concrete due to corrosion.

Static-load tests should be utilized only when the analytical method is impractical or otherwise unsatisfactory. The load test is the best tool in case of the structural element details are not readily available; deficiencies in details, materials, or construction methodology data and you find that the concrete strength is expected to be weak than the design concrete strength in this case the slab or bridge are best evaluated by a load test; and it can be used also if the design is extremely complex with limited prior experience for a structure of this type.

Inspection Methodology

Conditions for this test will be as follows:

1. results of a static load test permit a reasonable interpretation of structural adequacy;
2. principal structural elements under investigation are primarily flexural members;
3. adjacent structure's effects can be accounted for in the evaluation of the load test results.

Before conducting a load test, some repair actions can be required and an approximate analysis should be conducted. After establishing the magnitude of the test load, the load is applied incrementally with deflections measured. The structure is considered to have passed the load test if it shows no visible evidence of failure, such as excessive cracking or spalling, and it meets requirements for deflection. In certain applications, serviceability requirements, such as allowable leakage at maximum load, can also be a criterion.

Any viable design method or assessment of service life involves a number of essential elements: a behavioral model, acceptance criteria defining satisfactory performance, loads under which these criteria should be satisfied, relevant characteristic material properties, and factors or margins of safety that take into account uncertainties in the overall system as presented by Sommerville (1992).

The selection of materials and mixture proportions, such as the maximum w/c ratio, and structural detail considerations, provides one approach used for design of durable structures. Another approach entails prediction of service life using calculations based on knowledge about the current damage, degradation mechanisms, and the rates of degradation reactions. Development of a more comprehensive approach for design of durable structures requires integration of results obtained from a large number of studies that have been conducted relative to concrete durability

As discussed in the previous chapter, the main method of predicting the structure's reliability along its service life under future operating conditions is through probability-based techniques involving time-dependent reliability analyses. These techniques integrate information on design requirements, material and structural degradation, damage accumulation, environmental factors, and nondestructive evaluation technology into a decision tool that provides a quantitative measure of structural reliability.

In-service inspection methods can impact the structural reliability assessment in two areas: detection of defects and modifications to the frequency distribution of resistance.

In general, cracks in concrete have many causes. They may affect appearance only, or they may indicate significant structural distress or a lack of durability. Cracks may represent the total extent of the damage, or they may point to problems of greater magnitude. Their significance depends on the type of structure, as well as the nature of the cracking. For example, cracks that are acceptable for building structures may not be acceptable in water retaining wall structures.

The cracks may be happening in plastic concrete or hardened concrete. The sample of the plastic concrete cracks is plastic shrinkage cracking, settlement cracking and after hardening there are a dry shrinkage cracking will be illustrated as follows based on ACI code:

According to the Concrete Society Technical Report No. 54, different types of cracks occur at different times in the life of a concrete element, as shown in Table 8.1. So as well as a recognition of a crack pattern, a knowledge of the time of the first appearance of cracks is helpful in diagnosing the underlying cause.

8.2.3.1 Plastic Shrinkage Cracking

Cracking caused by plastic shrinkage in concrete occurs most commonly on the exposed surfaces of freshly placed floors and slabs or other elements with large surface areas when they are subjected to a very rapid loss of moisture caused by low humidity and wind or high temperature or both.

Plastic shrinkage usually occurs prior to final finishing, before curing start.

When moisture evaporates from the surface of freshly placed concrete faster than it is placed by bleed water, the surface concrete shrinks.

Due to the restrained provided by the concrete on the drying surface layer, tensile stresses develop in the weak, stiffening plastic concrete, resulting in shallow cracks that are usually not short and run in all directions.

In most cases, these cracks are wide at the surface. They range from a few millimeters to many meters in length and are spaced from a few centimeters to as much as 3 m apart. Plastic shrinkage cracks may extend the full depth of elevated structural slabs.

8.2.3.2 Settlement Cracking

After initial placement, vibration, and finishing, the concrete has a tendency to continue to consolidate or settle. During this period, the plastic concrete may be locally restrained by reinforcing steel, previously placed concrete, or formwork tie-bolts. These local restraints may result in voids and/or cracks adjacent to restraining the element as shown in Figure 8.1. When associated with reinforcing steel bars, this settlement cracking increases with increasing bar size, increasing slump, and decreasing cover. The degree of settlement cracking will be magnified by insufficient vibration or the use of leaking or highly flexible forms.

TABLE 8.1
Typical Times for Appearance of Defects

Type of Defect	Typical Time of Appearance
Plastic settlement cracks	10 min to 3 h
Plastic shrinkage cracks	30 min to 6 h
Crazing	1–7 days, sometimes much longer
Early thermal contraction cracks	1–2 days until 3 weeks
Long-term drying shrinkage cracks	Several weeks or months

Source: Concrete Society Technical Report 54.

Inspection Methodology

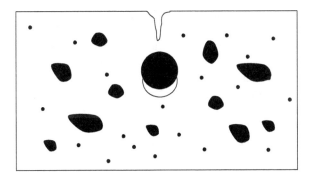

FIGURE 8.1 Settlement cracking.

Proper form design and adequate vibration, provision of sufficient time interval between the placement of concrete in slabs and beam, the use of the lowest possible slump, and an increase in concrete cover will reduce settlement cracking.

8.2.3.3 Drying Shrinkage

The common cause of cracking in concrete is drying shrinkage. This type of shrinkage is caused by the loss of moisture from the cement paste constituent, which can shrink by as much as 1% per unit length. Unfortunately, aggregate provides internal restraint that reduces the magnitude of this volume change to about 0.05%. On wetting, concrete tends to expand.

These moisture-induced volume changes are a characteristic of concrete. If the shrinkage of concrete could take place without any restraint, the concrete would not crack. It is the combination of shrinkage and restraint which is usually provided by another part of the structure or by the subgrade that causes tensile stresses to develop. When the tensile stresses of concrete is exceed, it will crack. Cracks may propagate at much lower stresses than are required to cause crack initiation.

In massive concrete structure elements as in the case of foundation under heavy machine, tensile stresses are used by differential shrinkage between the surface and the interior concrete. The larger shrinkage at the surface causes cracks to develop that may with time penetrate deeper into the concrete.

The following are the main factors that affect the magnitude of the tensile stresses:

1. Including the amount of shrinkage.
2. The degree of restraint.
3. The modulus of elasticity.
4. The amount of creep.

The amount of drying shrinkage is influenced mainly by the amount and type of aggregate and the water content of the mixture. The greater the amount of aggregate, the smaller the amount of shrinkage. The higher the stiffness of the aggregate, the more effective it is in reducing the shrinkage of the concrete, meaning that the concrete that contains sandstone aggregate has a higher shrinkage for about twice than that in case of concrete containing granite, basalt or limestone. The higher the water content, the greater the amount of drying shrinkage.

Surface crazing on walls and slabs is an excellent example of shrinkage due to drying on a small scale. Crazing usually occurs when the surface layer of the concrete has higher water content than that of the interior concrete. The result is a series of shallow, closely spaced fine cracks.

Shrinkage cracking can be controlled by using properly spaced contraction joints and proper steel detailing. Shrinkage cracking may also be controlled using shrinkage-compensating cement (Figure 8.2).

8.2.3.4 Thermal Stresses

The temperature differences within a concrete structure may be due to cement hydration or changes in ambient temperature conditions or both. These temperature differences result in differential volume changes. The concrete will crack, when the tensile strains exceed their tensile strain capacity due to the differential volume changes.

The effects of temperature differentials due to the hydration of cement are normally associated with mass concrete such as large columns, piers, beams, footing, retaining walls and dams, while temperature differentials due to changes in the ambient temperature can affect any structure.

The concrete rapidly gains both strength and stiffness as cooling begins. Any restraint of the free contraction during cooling will result in tensile stress. Tensile stresses developed during the cooling stage are proportional to the temperature change, the coefficient of thermal expansion, the effective modulus of elasticity, and the degree of restraint. The more massive the structure, the greater the potential for temperature differential and degree of restraint.

Procedures to help reduce thermally induced cracking include reducing the maximum internal temperature, controlling the rate at which the concrete cools, and increasing the tensile strain capacity of the concrete.

Hardened concrete has a coefficient of thermal expansion that may range from 7 to $11 \times 10^{-6}/°C$ with an average of $10 \times 10^{-6}/°C$. When one portion of a structure is subjected to a temperature induced volume change, the potential for thermally induced cracking exists.

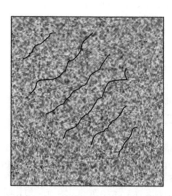

FIGURE 8.2 Drying shrinkage shown in concrete service.

Inspection Methodology

The designers must give special consideration to structures in which some portions are exposed to temperature changes, while other portions of the structure are either partially or completely protected. Figure 8.3 presents a water tank exposed directly to the sun from this the thermal effect will affect a tank cracks and deformation as the cylindrical tank convert to ellipse. A drop in temperature may result in the cracking of the exposed element, while increase in temperature may cause cracking in the protected portion of the structure. So the designer must allow movement of the structure by recommending the use of contraction joints and providing the correct and detailing to it.

These structures that have a high difference in temperature are usually built in areas near the desert, where there are more differences in temperature between afternoon and midnight. Moreover, in countries with high temperature, they usually use air condition inside the building so there will be a high probability to have cracks due to the difference in temperature inside and outside the building. So the designer should take these stresses into consideration.

As shown in Figure 8.4, the cracks usually appear between the masonry bricks wall and the concrete columns and beams special in case of external wall and this is due to different coefficient of variation between the brick and the concrete. So this should be considered in the structural and architectural detailed and specification to use a special joint materials to put in between the wall and the concrete frame.

Noting that high temperature and thermal gradients affect concrete's strength and stiffness. In addition, thermal exposure can result in cracking or, when the rate of heating is high and concrete permeability low, surface spalling can occur. Resistance of concrete to daily temperature fluctuations is provided by embedded steel reinforcement as described in ACI 318. There is a special cooling, to limit the concrete temperature to a maximum of 65C, except for local areas where temperatures can increase to 93C. At that temperature, there is the potential for DEF to occur if concrete is also exposed to moisture. These codes, however, do allow higher temperatures if tests have been performed to evaluate the strength reduction, and the design capacity is computed using the reduced strength. Because the response of concrete to elevated temperature is generally the result of moisture change effects, guidelines for

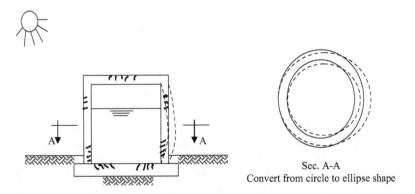

Sec. A-A
Convert from circle to ellipse shape

FIGURE 8.3 Water tank deformation due to thermal effect.

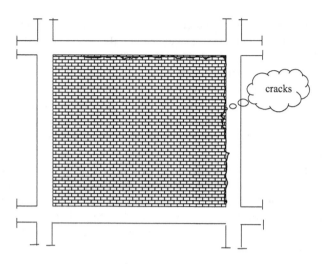

FIGURE 8.4 Cracks between masonry wall and the concrete members due to thermal effect.

development of temperature resistant reinforced concrete structures need to address factors, such as type and porosity of aggregate, permeability, moisture state, and rate of heating.

8.2.3.5 Alkaline Aggregate Reaction

Expansion and cracking leading to loss of strength, stiffness, and durability of concrete can result from chemical reactions involving alkali ions from Portland cement, calcium and hydroxyl ions, and certain siliceous constituents in aggregates. Expansive reactions can also occur as a result of interaction of alkali ions and carbonate constituents. Three requirements are necessary for disintegration due to alkali–aggregate reactions:

1. presence of sufficient alkali;
2. availability of moisture; and
3. the presence of reactive silica, silicate, or carbonate aggregates.

Controlling alkali–aggregate reactions at the design stage is done by avoiding deleteriously reactive aggregate materials by using preliminary petrographic examinations and by using materials with proven service histories. ASTM C 586 provides a method for assessing potential alkali reactivity of carbonate aggregates. ACI 201.2R presents a list of known deleteriously reactive aggregate materials (Table 8.2).

Additional procedures for mitigating alkali–silica reactions include pozzolans, using low-alkali cements (i.e., restricting the cement alkali contents to less than 0.6% by weight sodium oxide [Na-O] equivalent), adding lithium salts, and applying barriers to restrict or eliminate moisture. The latter procedure is generally the first step in addressing affected structures. The alkali–carbonate reaction can be controlled by keeping the alkali content of the cement low, by adding lithium salts, or by diluting the reactive aggregate with less-susceptible material (Figure 8.5).

Inspection Methodology

TABLE 8.2
Effect of Commonly Used Chemicals on Concrete

Different Substances Attack Concrete					Rate of Attack at Ambient Temperature
Inorganic Acids	**Organic Acids**	**Alkaline Solution**	**Salt Solutions**	**Miscellaneous**	
Hydrochloric Nitric Sulfuric	Acetic Formic Lactic	–	Aluminum chloride	–	Rapid
Phosphoric	Tannic	Sodium or potassium hydroxide >20%	Ammonium nitrate Ammonium sulfate Sodium sulfate Magnesium sulfate Calcium sulfate	Bromine (gas) Sulfate (liquor)	Moderate
Carbonic	–	Sodium or potassium hydroxide 10%–20%	Ammonium chloride Magnesium chloride Sodium cyanide	Chlorine (gas) Seawater Soft water	Slow
–	Oxalic Tartaric	Sodium or potassium hydroxide <10% Sodium chloride Ammonium hydroxide	Calcium chloride Sodium chloride Zinc nitrate Sodium chromate	Ammonia (liquid)	Negligible

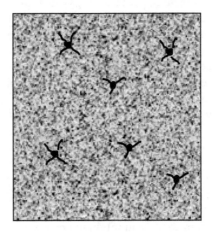

FIGURE 8.5 Aggregates with alkaline.

As shown in Figure 8.6, cracks like a star in the concrete surface are an indication of a chemical reaction. This reaction occurs if there are some aggregates containing active silica and alkalis derived from cement hydration, admixtures, or external sources such as curing water, ground water, alkaline solutions stored or used in the finished structure.

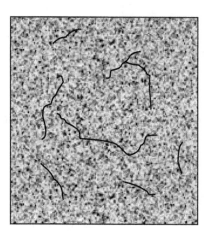

FIGURE 8.6 Result of sulfate salt.

The alkali–silica reaction results in the formation of a swelling gel, which tends to draw water from other portions of the concrete. This causes local expansion and accompanying tensile stresses, and may eventually result in the complete deterioration of the structure.

8.2.3.6 Sulfate Attack

This usually happens to a foundation because it is in contact with soils, groundwater, and seawater, which react with the calcium hydroxide $Ca(OH)_2$ and the hydrated tricalcium aluminate (C_3A) to form gypsum and ettringite, respectively. These reactions can result in deleterious expansion and produce concretes with reduced strength because of decomposition and expansion of the hydrated calcium aluminates.

Increased resistance of structures to sulfate attack is provided by fabricating them using concrete that is dense, has low permeability, and incorporates sulfate-resistant cement. Because it is the C_3A that is attacked by sulfates, the concrete vulnerability can be reduced by using cements low in C_3A, such as ASTM C 150 Types II and V sulfate-resisting cements. Under extreme conditions, supersulfated slag cements such as ASTM C 595 Types VP or VS can be used. Also, improved sulfate resistance can be attained by using admixtures, such as pozzolans and blast-furnace slag. The requirements and guidelines for the use of sulfate-resistant concretes are based on exposure severity and are provided in ACI 318 and ACI 201.2R. The requirements are provided in terms of cement type, cement content, maximum w/cm, and minimum compressive strength, depending upon the potential for distress.

The ground water that has sulfate is a special durability problem for concrete. Due to the sulfate penetrates hydrated cement paste, it comes in contact with hydrated calcium aluminates. Calcium sulfoaluminate will be formed, accompanied by an increase in the volume resulting in high tensile stresses causing cracking. Therefore, it is usually recommend using Portland cement type ASTM C150

Inspection Methodology

II and V, which contain low tricalcuim aluminates that will reduce the severity of the cracks.

8.2.4 ENVIRONMENTAL CONSIDERATIONS

Design of reinforced concrete structures to ensure adequate durability is a complicated process (see Chapter 8 for more details). Service life depends on structural design and detailing, mixture proportioning, concrete production and placement, construction methods, and maintenance. Also, changes in use, loading, and environment are important. Because water or some other fluid is involved in almost every form of concrete degradation, concrete permeability is important.

The process of chemical and physical deterioration of concrete with time or reduction in durability is generally dependent on the presence and transport of chloride, carbonate or sulfate substances through concrete, and the magnitude, frequency, and effect of applied loads.

The combined transportation of heat, moisture, and chemicals, both within the concrete and in exchange with the surrounding environment, and the parameters controlling the transport mechanisms constitute the principal elements of durability. The rate, extent, and effect of fluid transport are largely dependent on the concrete pore size and distribution, presence of cracks, and microclimate at the concrete surface. The primary mode of transport in uncracked concrete is through the bulk cement paste pore structure and the transition zone which is the interfacial region between the particles of coarse aggregate and hydrated cement paste.

The physical–chemical phenomena associated with fluid movement through porous solids are controlled by the concrete permeability. Although the coefficient of permeability of concrete depends primarily on the w/cm and maximum aggregate size, it is also influenced by age, consolidation, curing temperature, drying, and the addition of chemical or mineral admixtures.

Based on Mehta (1986), the concrete is generally more permeable than cement paste due to the presence of microcracks in the transition zone between the cement paste and aggregate.

As an example, the permeability is affected by the concrete mix properties as shown in Table 8.3. The chloride diffusion and permeability results obtained from the 19 mm maximum size crushed limestone aggregate. This test was performed for six different concrete mixtures as presented in Table 8.4.

Two additional factors are considered with respect to fabrication of durable concrete structures: the environmental exposure condition and specific design recommendations pertaining to the expected form of aggressive chemical or physical attack (e.g., designing the structure to prevent accumulation of water). Exposure conditions or severity are generally handled through a specification that addresses the concrete mixture (e.g., strength, w/c, and cement content), and details (such as concrete cover), as dictated by the anticipated exposure.

Mixture 1 is expressed as the ratio of water to total cementitious material content at these degradation mechanisms. Combined effects where more than one of these processes can be simultaneously occurring are also briefly addressed.

TABLE 8.3
Chloride Transport and Permeability Results for Selected Concretes Based on Whiting, 1988

Mixture No.	Cure Time, days	% Cl-by Weight of Concrete	Permeability, 10^{-6} m^2/s Hydraulic	Air	Porosity, % by Volume
1	1	0.013		37	8.3
	7	0.013		29	7.5
2	1	0.017		28	9.1
	7	0.022		33	8.8
3	1	0.062	0.030	130	11.3
	7	0.058	0.027	120	11.3
4	1	0.103	0.560	120	12.4
	7	0.076	0.200	170	12.5
5	1	0.104	0.740	200	13.0
	7	0.077	0.230	150	12.7
6	1	0.112	4.100	270	13.0
	7	0.085	0.860	150	13.0

TABLE 8.4
Concrete Mixture Proportions and Characteristics

Mixture No.	Quantities, kg/m³ Cement	Fine Aggregate	Coarse Aggregate	Water	Admixture(s)	w/c	Slump, cm	Air Content, %
1	446	752	1032	132	A+B	0.258	119	1.6
2	446	790	1083	128	C	0.288	89	2.0
3	381	784	1075	153	D	0.401	89	2.3
4	327	794	1088	164	–	0.502	94	2.1
5	297	791	1086	178	–	0.600	107	1.8
6	245	810	1107	185	–	0.753	124	1.3

Note: A is Microsilica fume at 59.4 kg/m³; B is Type F high-range water reducer at 25 ml/kg; C is Type F high-range water reducer at 13 ml/kg; and D is Type A water reducer at 2 ml/kg.

Available methods and strategies for prediction of the service life of a new or existing reinforced concrete structure with respect to these mechanisms are described in Chapter 8.

8.2.4.1 Chemical Attack

Chemical attack involves the alteration of concrete through chemical reaction with either the cement paste, coarse aggregate, or embedded steel reinforcement. Generally, the attack occurs on the exposed surface region of the concrete (cover concrete), but with the presence of cracks or prolonged exposure, chemical attack can affect entire structural cross sections. Chemical causes of deterioration can be grouped into three categories (Mehta, 1986):

1. Hydrolysis of cement paste components by soft water;
2. Cation–exchange reactions between aggressive fluids and cement paste; and
3. Reactions leading to formation of expansion product. Results from prolonged chemical attack range from cosmetic damage to loss of structural section and monolithic behavior. Chemical attack of embedded steel reinforcement can also occur.

8.2.4.2 Leaching

This phenomenon will occur if the pure water that contains little or no calcium ions, or acidic ground water present in the form of dissolved carbon dioxide gas, carbonic acid, or bicarbonate ion, tends to hydrolyze or dissolve the alkali oxides and calcium-containing products resulting in increasing permeability. The rate of leaching is dependent on the amount of dissolved salts contained in the percolating fluid, rate of permeation of the fluid through the cement paste matrix, and temperature.

Note that the rate of leaching can be lowered by minimizing the permeation of water through the interconnected capillary cavities into the concrete by using low-permeability concretes or using a coating as physical barriers.

Factors related to the production flow-permeability concretes include low *w/c*, adequate cement content, and proper compaction and curing conditions. Polymeric modification can also be used to provide low permeability concretes. Similarly, attention should be given to aggregate size and gradation, thermal and drying shrinkage strains, avoiding loads that produce cracks, and designing and detailing to minimize exposure to moisture. Requirements in codes and suggested guidelines for *w/c ratio* are generally based on strength or exposure conditions.

8.2.4.3 Acid and Base Attack

Acids can combine with the calcium compounds in the hydrated cement paste to form soluble materials that are readily leached from the concrete to increase porosity and permeability. The main factors determining the extent of attack are type of acid, and its concentration and pH. Protective barriers are recommended to provide resistance against acid attack.

As hydrated cement paste is an alkaline material, concrete made with chemically stable aggregates is resistant to bases. Sodium and potassium hydroxides in high concentrations (>20%), however, can cause concrete to disintegrate. ACI 515.1R provides a list of the effects of chemicals on concrete. Under mild chemical attack, a concrete with low *w/c* (low permeability) can have suitable resistance. Because corrosive chemicals can attack concrete only in the presence of water, designs to minimize attack by bases might also incorporate protective barrier systems. Guidelines on the use of barrier systems are also provided in ACI 515.1R.

8.2.4.4 Steel Reinforcement Corrosion

Corrosion of conventional steel reinforcement in concrete is an electrochemical process that forms either local pitting or general surface corrosion. Both water and oxygen must be present for corrosion to occur. In concrete, reinforcing steel with adequate cover should not be susceptible to corrosion because the highly alkaline conditions present within the concrete (pH > 12) cause a passive iron-oxide film to form on the steel surface. Carbonation and the presence of chloride ions, however, can destroy the protective film. Corrosion of steel reinforcement also can be accelerated by the presence of stray electrical currents.

As noted in the following figures, in case of corrosion in the steel reinforcement, the cracks you will see will be parallel to the steel bars for any concrete members such as beams, slabs, and columns.

Figure 8.7 presents the shape of cracks in case of slab or side of the foundation. Figure 8.8 shows the corrosion of steel bars for slab after removing the concrete cover.

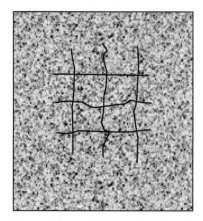

FIGURE 8.7 Cracks due to steel corrosion (cracks parallel to steel bars).

FIGURE 8.8 Corrosion of steel bars in slab.

Inspection Methodology

The bottom of beam with corrosion in main steel bars is shown in Figure 8.9 while present the corrosion of steel bars on the column in Figure 8.10.

Moreover, spots of brown color on the concrete surface shows an indication of the corrosion in the steel reinforcement in this stage. Only the qualified engineers should perform the inspection.

The disadvantage of the visual inspection is that it completely depends on the experience of the engineer who performs the inspection since some cracks have more than one reason with or without the corrosion effect.

Figure 8.11 presents the corrosion on the longitudinal steel bars for the foundation underneath the legs of the sphere tank.

Figure 8.12 presents a combined deterioration of the foundation for many reasons as the corrosion of the main steel and the thermal expansion of concrete and steel are different between concrete and grouting.

Figure 8.13 illustrates the corrosion in steel foundation under steel structure. There is a deficiency in the grouting and also signs of longitudinal cracks on the

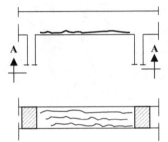

Sec. A-A Cracks parallel to main steel in case of corrosion

FIGURE 8.9 Cracks due to corrosion or concrete covered not enough.

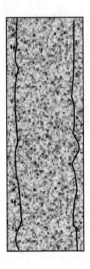

FIGURE 8.10 Cracks due to corrosion.

FIGURE 8.11 Corrosion in steel foundation under sphere tanks.

FIGURE 8.12 Corrosion in steel foundation under pipe support.

FIGURE 8.13 Corrosion in steel foundation under steel structure.

corner mean the corrosion has started. The longitudinal cracks as there are some lifting force and weak in concrete strength.

Figure 8.14 presents cracks on the foundation underneath the static equipment. The cracks on the grouting due to weather effect and high temperature in most cases and these cracks is normal for cement grout. The longitudinal cracks due to increase in loads and weak on the concrete strength or design deficiency.

Inspection Methodology

FIGURE 8.14 Corrosion in foundation under mechanical skid.

The penetration of carbon dioxide (CO_2) from the environment decreases the pH of concrete as calcium and alkali hydroxides are converted into carbonates. The penetration of CO_2 generally is a slow process, dependent on the concrete permeability, the concrete moisture content, the CO_2 content, and ambient relative humidity (RH). Carbonation can be accelerated by the presence of cracks or porosity of the concrete. Concretes that have low permeability and have been properly cured provide the greatest resistance to carbonation. Also, concrete cover over the embedded steel reinforcement can be increased to delay the onset of corrosion resulting from the effects of carbonation.

The presence of chloride ions is probably the major cause of corrosion of embedded steel reinforcement. Chloride ions are common in nature and small amounts can be unintentionally contained in the concrete mixture ingredients. Potential external sources of chlorides include those from accelerating admixtures (e.g., calcium chloride), application of deicing salts, or exposure to seawater or spray. Maximum permissible chloride-ion contents, as well as minimum concrete cover requirements, are provided in codes and guides ACI 318, EC2 and BS8110.

The following two methods are the most commonly used methods for determining the chloride contents in concrete: acid-soluble test (total chlorides) and water-soluble test. The chloride ion limits are presented in terms of type of member prestressed or conventionally reinforced) and exposure condition for dry or moist. The concrete permeability plays a key role in controlling the process because of water, oxygen, and chloride ions which are important factors in the corrosion of embedded steel reinforcement. Concrete mixtures should be designed to ensure low permeability by using low *w/c ratio* adequate cementitious materials contents, proper aggregate size and gradation, and mineral admixtures.

Methods of excluding external sources of chloride ions from existing concrete, as detailed in ACI 222R, include using waterproof membranes, polymer impregnation, and overlay materials.

For evaluating the new or old building to predict its service life, it is very important to know if the structure uses some protection for corrosion, such as cathodic protection system, use of stainless steel bars or epoxy coated bars, or use of admixtures as corrosion inhibitors or any other protection.

Prestressed concrete is usually a form of concrete used in construction of bridges, which is prestressed be being placed under compression prior to supporting any loads beyond its own dead weight. Therefore, its reliability is very sensitive to any reduction on steel section.

The corrosion of prestressing steel can be either highly localized by pitting or uniform corrosion. In most cases, much pitting combines to be a uniform corrosion, which is a result of localized attack resulting in pitting, stress corrosion, hydrogen embrittlement, or a combination of these. Pitting is an electrochemical process that results in local penetrations into the steel to reduce the cross-section, so that it is incapable of supporting its load. Stress-corrosion cracking results in the brittle fracture of a normally ductile metal or alloy under stress in specific corrosive environments. Hydrogen embrittlement, frequently associated with exposure to hydrogen sulfide, occurs when hydrogen atoms enter the metal lattice and significantly reduce its ductility. Hydrogen embrittlement can also occur as a result of improper application of cathodic protection to the post-tensioning system. Due to the magnitude of the load in the posttensioning systems, the tolerance for a corrosion attack is less than for mild steel reinforcement. Corrosion protection is provided at installation by either encapsulating the post-tensioning steel with microcrystalline waxes compounded with organic corrosion inhibitors within plastic sheaths or metal conduits (unbounded tendons), or by Portland cement (grouted tendons).

8.2.4.5 Salt Crystallization

Salts can produce cracks in concrete through development of crystal growth pressures that arise from causes, such as repeated crystallization due to evaporation of salt-laden water in the pores. Structures in contact with fluctuating water levels or in contact with ground water containing large quantities of dissolved salts which are calcium sulfate ($CaSO_4$), sodium chloride (NaCl), sodium sulfate (Na_2SO_4) are susceptible to this type of degradation, in addition to possible chemical attack, either directly or by reaction with cement or aggregate constituents. One approach to the problem of salt crystallization is to apply sealers or barriers to either prevent water ingress or subsequent evaporation; however, if the sealer is not properly selected and applied, it can cause the moisture content in the concrete to increase, and not prevent the occurrence of crystallization.

8.2.4.6 Freezing-and-Thawing Attack-Concrete

In the countries that have a cold climate, the concrete is exposed to freezing and thawing cycles, and also this happens in special industries. The major deterioration resulting from these phenomena is the freezing of the water inside the concrete pore due to the capillary rise. So, the deterioration is a result of a hydraulic pressure of the freezing water.

There are some guides (e.g., ACI201.2R and ACI318 by) to designing a durable concrete structure under freezing and thawing cycle, which define the total air content that is related to the maximum aggregate size and exposure condition, as well as the recommended w/c ratio as a function of the concrete cover, and presence of aggressive agents as deicing chemicals.

Factors controlling the resistance of concrete to freezing-and-thawing action include air entrainment (size and spacing of air voids), permeability, strength, and degree of saturation. Because the degree of saturation is important, concrete structures should be designed and detailed to promote good drainage. ASTM C 682 provides a test that allows the user to specify the curing history of the specimen and the exposure conditions that most nearly match the expected service conditions. An estimate of the susceptibility of concrete aggregates for known or assumed field environmental conditions should be considered. Structures constructed without adequate air entrainment can have an increased risk for freezing-and-thawing damage.

8.2.4.7 Abrasion, Erosion, and Cavitation

Abrasion, erosion, and cavitation of concrete results in progressive loss of surface material. This is usually happens in special concrete member such as a concrete deck for parking garage, bridges, slab on grade in the workshop. In some cases, the tank that contains a fluctuating liquid with small particle and the erosion is usually seen on the hydraulic structure.

Abrasion generally involves dry attrition, while erosion involves a fluid containing solid particles in suspension. Cavitation causes loss of surface material through the formation of vapor bubbles and their sudden collapse.

The abrasion and erosion resistance of concrete is affected primarily by the strength of the cement paste, the abrasion resistance of the fine and coarse aggregate materials, and finishing and curing. Special toppings, such as coats of cement and iron aggregate on the concrete surface, can be used to increase abrasion resistance. Therefore, the design for durable structure that exposed to abrasion and erosion should specify clearly the abrasion test for aggregate and also special are a suitable coating materials.

Concrete that resists abrasion and erosion can still suffer severe loss of surface material due to cavitation. The best way to guard against the effects of cavitation is to eliminate its cause(s).

8.2.4.8 Combined Effects

Degradation of concrete, particularly in its advanced stages, is seldom due to a single mechanism. The chemical and physical causes of degradation are generally so intertwined that separating the cause from the effect often becomes impossible (Mehta, 1986). Limited information is available relative to the assessment of the remaining service life of concrete exposed to the combined effects of freezing-and-thawing degradation (surface scaling) and corrosion of steel reinforcement (Fagerlund et al., 1994) (Figures 8.15 and 8.16).

8.3 STEEL STRUCTURE INSPECTION

8.3.1 VISUAL TEST

Some precautions that should be followed when doing the visual inspection have been asserted by ASME V. Direct visual examination is usually made when access is sufficient to place the eye within 24 inches (600 mm) of the surface to be examined

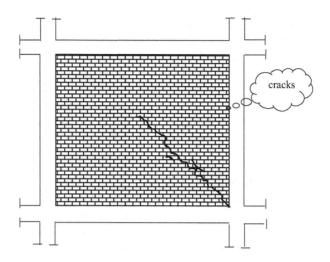

FIGURE 8.15 Inclination of the cracks in masonry wall due to settlement.

FIGURE 8.16 Ring beam failure

and at an angle not less than 30° to the surface to be examined. Mirrors may be used to improve the angle of vision, and aids, such as a magnifying lens, may be used to assist examinations.

Illumination (natural or supplemental white light) for the specific part, component, vessel, or section thereof being examined is required. The minimum light intensity at the examination surface/site shall be 100 foot candles (1,000 lux).

The light source, technique used, and light level verification needs to be demonstrated one time, documented, and maintained on file.

The visual inspection is usually performed by eye with the experience of the inspector so it depends mainly on the capability of the fitness of the inspector in addition to his long experience on welding. In addition to that, there are some tools

Inspection Methodology

that can be used on the site to check if the welding is up to the standard and project specification or not.

As shown in Figure 8.17, the fillet weld is a rectangular piece of metal on the four corners and it has the shape of the fillet weld. So the inspector will move this piece over the piece of the steel if the welding thickness matches with the value of the gage and it will be matched with the required thickness.

The other tools are used to measure the fillet weld for both the concavity and convexity filet weld shape, as illustrated in Figure 8.18. In addition, this tool is used to measure the weld reinforcement.

The Vernier caliper is an essential tool to measure the thickness of the base metal and weld with good accuracy, as shown in Figure 8.19.

The Vernier caliper is an extremely precise measuring instrument: the reading error is 1/20 mm = 0.05 mm.

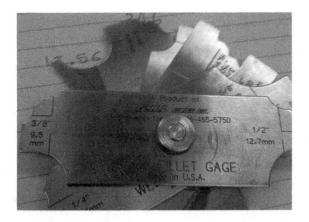

FIGURE 8.17 Fillet weld thickness test.

FIGURE 8.18 Fillet weld thickness.

FIGURE 8.19 Caliper Vernier.

This equipment is used if there is a pitting in the base metal due to corrosion or between the base metal and the weld due to poor welding process as the undercut and the pit measurement tools as shown in Figure 8.20.

These tools, for example, can be used to measure the angle of preparation, excess weld metal, depth of undercut, depth of pitting, weld throat size, and more, in Figure 8.21.

8.4 OFFSHORE STRUCTURE

The main inspection tools used in offshore structure are visual inspection and measuring thickness tool by UT inspection as the most common problem is the deterioration due to corrosion. The corrosion of the tubular section on the deck is presented in Figures 8.22 and 8.23.

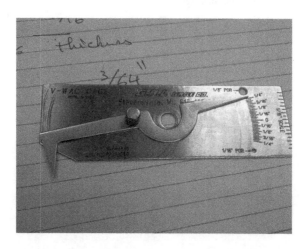

FIGURE 8.20 Steel pit measurement tool.

Inspection Methodology 205

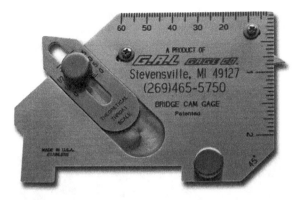

FIGURE 8.21 Multifunction tool for welding.

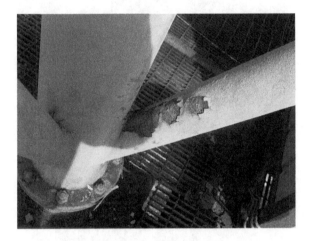

FIGURE 8.22 The corrosion on the tubular joint for the deck structure.

FIGURE 8.23 Corrosion of a truss member on a helideck.

Bridges which connect two platforms have a special support design as shown in Figure 8.24. The corrosion of bridges is very important as its repair is very expensive, and it shall be sliding so with corrosion and the roughness will increase with time which converts the release of the support moving as to convert it from roller to hinge which increases the stresses on the bridge members.

The corrosion can create a big hole like this for the leg for one of the flare jackets, as shown in Figure 8.25.

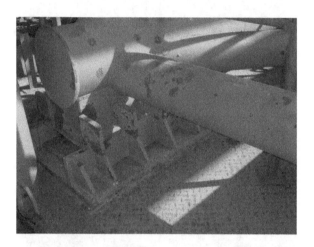

FIGURE 8.24 Corrosion of a bridge support.

FIGURE 8.25 Leg corrosion.

Inspection Methodology

Figure 8.26 presents a very unique case study, as by visual inspection only it seemed that there is a corrosion, but when starting the painting process implementation which is started by sand blasting a lot of holes are found in the tubular truss member for the helideck. So it is very important to define the condition of the structure before starting any maintenance. It is also important to note that this helideck is working under very high risk as the load of the helicopter is the same but the strength of the structure is very weak. So the main question why it has not failed? The answer is covered in Chapters 5 and 6.

Figure 8.27 presents the cutting member on the joint due to fatigue effect noting that this joint failure was a paper of other damage which causes the platform failure.

The underwater inspection accuracy is very important as it will be included in the structure model to perform pushover analysis to evaluate the offshore structure.

FIGURE 8.26 Truss corrosion.

FIGURE 8.27 Cutting joint.

BIBLIOGRAPHY

AASHTO T260-84, Standard method of sampling and testing for total chloride ion in concrete ratio materials, American Association of State Highway Transportation Officers, Washington, DC, 1984.
ACI 228-89-R1, In-place methods for determination of strength of concrete.
ACI 318-97, Standard building code requirements for reinforced concrete, Detroit, MI, 1997.
Alldred, J.C., Quantifying the losses in cover-meter accuracy due to congestion of reinforcement, *Proceedings of the Fifth International Conference on structural Faults and Repair*, Vol. 2, Engineering Techniques Press, Edinburgh, pp. 125–30, 1993.
American Concrete Institute, Corrosion of metals in concrete, Report by ACI Committee 222. ACI222R-89, American Concrete Institute, Detroit, MI, 1990.
ASTM C1152, Test method for acid-soluble chloride in mortar and concrete.
ASTM C1202, Test method for electrical indication of concrete's ability to resist chloride.
ASTM C341, Test method for length change of drilled or sawed specimens of hydraulic-cement mortar and concrete.
ASTM C42-90m, Standard test method for obtaining strength and testing drilled cores and sawed beams of concrete.
ASTM C597, Standard test method for pulse velocity through concrete.
ASTM C682, Standard test method for evaluation of frost resistance of coarse aggregates in air-entrained concrete by critical dilation procedures.
ASTM C805, Test method for rebound number in concrete.
ASTM C876, Standard test method for half cell potentials of reinforcing steel in concrete.
ASTM C876, Standard test method for half-cell potentials of uncoated reinforcing steel in concrete, American Society for Testing and Materials, Philadelphia, PA, 1991.
ASTM D1411-82, Standard test methods for water soluble chlorides present as admixes in graded aggregate road mixes, American Society for Testing and Materials, Philadelphia, PA, 1982.
Avram, C., 1981, *Concrete Strength and Strain*, Elsevier Scientific Publishing Co., New York.
British Standard Institution-BS 1881-1983-Part 120. Method for determination of the compressive strength of concrete cores.
Broomfield, J.P., Langford, P.E., and McAnoy, R., Cathodic protection for reinforced concrete: its application to buildings and marine structures, in Corrosion of Metals in Concrete, *Proceedings of Corrosion/87 Symposium*, Paper 142, NACE, Houston, TX, pp. 222–325, 1987.
Broomfield, J.P., Rodriguez, J., Ortega, L.M., and Garcia, A.M., Corrosion rate measurement and life prediction for reinforced concrete structures, *Proceeding of Structural Faults and Repair* – 93, Vol. 2, Engineering Technical Press, University of Edinburgh, Edinburgh, pp. 155–64, 1993.
BS 1881, Testing concrete: Part 5: Methods for testing hardened concrete for other than strength.
BS 1881, Part 6-1971, methods of testing concrete, analysis of hardened concrete.
BS 8110, structural user of concrete-Part 1:1985 Code of practice for design and construction, London.
Bungey, J.H., (ed.), Non-destructive testing in civil engineering, *International Conference by the British Institute of Non-destructive Testing*, Liverpool University, Liverpool, 1993.
Concrete Society, Concrete core testing for strength cement and concrete association, Wexham Spring, Slough SL3 6PL, Technical Report No.1 including addendum, 1987.
Concrete Society, Repair of concrete damaged by reinforcement corrosion, Technical Report No. 26, 1984.
Concrete Society, Diagnosis of deterioration in concrete structures, Technical Report No. 54, 2000, Concrete Society, Crowthorne.

El-Reedy, M.A., 2009, *Advanced Materials and Techniques for Reinforced Concrete Structure*, CRC Press, Boca Raton, FL.

Gerwick, Jr., B.C., 1981, High-amplitude low-cycle fatigue in concrete sea structures, *PCI Journal*, Chicago, IL, Sept.–Oct.

Hydro's Experience, Concrete durability, *Proceedings of the Katherine and Bryant Mather International Symposium*, SP-100, J. M. Scanlon, ed., American Concrete Institute, Farmington Hills, MI, pp. 1121–54, 2002.

Jacob, F., 1965, Lessons from failures of concrete structures, *Monograph* No. 1, American Concrete Institute, Farmington Hills, MI.

Kropp, J., and Hilsdorf, H.K., 1995, Performance criteria for concrete durability, TC-116-PCD, International Union of Testing and Research Laboratories for Materials and Structures (RILEM), E&FN Spon, Cachan Cedex, France.

Malhotra, V.M., and Carino, N.J., 1991, *Handbook of Nondestructive Testing of Concrete*, CRC Press, Boca Raton, FL.

Malhotra, V.M., ed., 1984, *In Situ/Nondestructive Testing of Concrete*, SP-82, American Concrete Institute, Farmington Hills, MI.

Murphy, W.E., 1984, Interpretation of tests on strength of concrete in structures, In: *Situ/Nondestructive Testing of Concrete*, SP-82, V.M. Malhotra, (ed.), American Concrete Institute, Farmington Hills, MI, pp. 377–92.

Naus, D.J., and Oland, C.B., 1994, Structural aging program technical progress report for period Jan. 1, 1993, to June 30, 1994, ORNL/NRC/LTR-94/21, Martin Marietta Energy Systems, Oak Ridge National Laboratory, Oak Ridge, Tenn., Nov.

Neville, A., 1991, *Properties of Concrete*, John Wiley & Sons, Inc., New York.

Parrott, L.J., A review of carbonation in reinforced concrete, a review carried out by C&CA under a BRE contract, British Cement Association, Slough, UK, 1987.

Price, W.H., 1951, Factors influencing concrete strength, *ACI Journal*, Vol. 47, No. 2, pp. 417–32.

Rewerts, T.L., 1985, Safety requirements and evaluation of existing buildings, *Concrete International*, Vol. 7, No. 4, pp. 50–55.

Sturrup, V.R., and Clendenning, T.G., 1969, The evaluation of concrete by outdoor exposure, Highway Research Record HRR-268, Washington, DC.

Sturrup, V.R., Hooton, R., Mukherjee, P., and Carmichael, T., 1987, Evaluation and prediction of durability—Ontario.

Woods, H., 1968, Durability of concrete construction, *ACI Monograph* No. 4, American Concrete Institute, Farmington Hills, MI.

9 Inspection Methods

9.1 CONCRETE STRUCTURES

After pouring the concrete, it will become in hardening shape and then we may need to conduct some tests and measurements to ascertain the quality of concrete. In some cases it is required to do non destructive tests in the case of a disagreement between the contractor and the owner or the consultant or in case of lack of confidence about the concrete specification to be matched with project specification requirement.

In that case, the quality control inspector should be able to deal with different ways to detect the quality of hardened concrete through nondestructive testing.

Therefore, choose the appropriate method that is economically feasible as well as from a structure condition and system point of view. The tests should be based on the condition of the structure and the location of the member that will be tested to be structurally safe, and the test should provide reasonable confidence and accuracy to the results (Table 9.1).

9.1.1 Core Test

This test is considered to be one of the semi-destructive tests. This test is very important and popular for the study of the safety of a structure as a result of changing the system of loading, or a deterioration as a result of fire or weather factors, or the need for temporary support for repair when there are no any accurate data about concrete strength.

This test is not too expensive and, on the other hand, it is the most accurate test to know the strength of concrete actually carried out.

It is worth mentioning that core testing is done through cutting of cylinders from concrete member, which could affect the integrity of structure and therefore the required samples must be taken according to the standard as the required number will provide us adequate accuracy of the results without weakening the building without benefit.

In our case, deterioration due to corrosion of steel bars causes the structure to lose most of its strength. More caution must be taken when performing this test by selecting the proper concrete member, which will not affect at all the building from structure point of view.

Therefore, the codes and specifications provide some guidance to the number of cores to be tested and these values are as follows:

- Volume of concrete member (V) $\leq 150\,m^3$ takes 3 cores
- Volume of concrete member (V) $> 150\,m^3$ takes $(3 + (V - 150 / 50))$ cores

TABLE 9.1
Nondestructive Test Methods for Determining Material Properties of Hardened Concrete and Assess the Condition in Existing Construction (ACI 228.2)

	Possible Methods		
Property	Primary	Secondary	Comment
Compressive strength	Cores for compression testing (ASTM C42 and C39)	Penetration resistance (ASTM C803; pullout testing drilled in)	Strength of in-place concrete; comparison of strength in different locations; and drilled-in pullout test not standardized
Relative compressive strength	Rebound number (ASTM C805); ultrasonic pulse velocity (ASTM C597)	–	Rebound number influenced by near surface properties; ultrasonic pulse velocity gives average result through thickness
Tensile strength	Splitting-tensile strength of core (ASTM C496)	In-place pulloff test (ACI 503R; BS 1881; Part 207)	Assess tensile strength of concrete
Density	Specific gravity of samples (ASTM C642)	Nuclear gage	–
Moisture content	Moisture meters	Nuclear gage	–
Static modulus of elasticity	Compression test of cores (ASTM C469)	–	–
Dynamic modulus of elasticity	Resonant frequency testing of sawed specimens (ASTM C215)	Ultrasonic pulse velocity (ASTM C597); impact echo; spectral analysis of surface waves (SASW)	Requires knowledge of density and Poisson's ratio (except ASTM C215); dynamic elastic modulus is typically greater than the static elastic modulus
Shrinkage/expansion	Length change of drilled or sawed specimens (ASTM C341)	–	Measure of incremental potential length change
Resistance to chloride penetration	90-day ponding test (AASHTO-T-259)	Electrical indication of concrete's ability to resist chloride ion penetration (ASTM C1202)	Establishes relative susceptibility of concrete to chloride ion intrusion; assess effectiveness of chemical sealers, membranes, and overlays
Air content; cement content; and aggregate properties (scaling, alkali-aggregate reactivity, freezing-and-thawing susceptibility	Petrographic examination of concrete samples removed from structure (ASTM C856, ASTM C457); Cement content (ASTM C1084)	Petrographic examination of aggregates (ASTM C294, ASTM C295)	Assist in determination of cause(s) of distress; degree of damage; quality of concrete when originally cast and current

(Continued)

TABLE 9.1 (*Continued*)
Nondestructive Test Methods for Determining Material Properties of Hardened Concrete and Assess the Condition in Existing Construction (ACI 228.2)

	Possible Methods		
Property	**Primary**	**Secondary**	**Comment**
Alkali-silica reactivity	Cornel VSHRP rapid test	–	Establish in field if observed deterioration is due to alkali-silica reactivity
Carbonation, pH	Phenolphthalein (qualitative indication); pH meter	Other pH indicators (e.g., litmus paper)	Assess corrosion protection value of concrete with depth and susceptibility of steel reinforcement to corrosion; depth of carbonation
Fire damage	Petrography; rebound number (ASTM C805)	Ultrasonic pulse velocity; impact-echo; impulse response	Rebound number permits demarcation of damaged concrete
Freezing-and-thawing damage	Petrography	SASW; impulse response	–
Chloride ion content	Acid-soluble (ASTM C1152) and water-soluble (ASTM C1218)	Specific ion probe	Chloride ingress increases susceptibility of steel reinforcement to corrosion
Air permeability	SHRP surface airflow method (SHRP-S-329)	–	Measures in-place permeability index of near surface concrete (15 mm)
Electrical resistance of concrete	AC resistance using four-probe resistance meter	SHRP surface resistance test (SHRP-S-327)	AC resistance useful for evaluating effectiveness of admixtures and cementitious additions; SHRP method useful for evaluating effectiveness of sealers
Reinforcement location	Covermeter; ground penetrating radar (GPR) (ASTM D4748)	X-ray and Y-ray radiography	Steel location and distribution; concrete cover
Local or global strength and behavior	Load test, deflection or strain measurements	Acceleration, strain, and displacement measurements	Ascertain acceptability without repair or strengthening; determine accurate load rating
Corrosion potentials	Half-cell potential (ASTM C876)	–	Identification of location of active reinforcement corrosion
Corrosion rate	Linear polarization (SHRP-S-324 and S-330)	–	Corrosion rate of embedded steel; rate influenced by environmental conditions

(*Continued*)

TABLE 9.1 (*Continued*)
Nondestructive Test Methods for Determining Material Properties of Hardened Concrete and Assess the Condition in Existing Construction (ACI 228.2)

Property	Possible Methods Primary	Secondary	Comment
Locations of delamination, voids, and other hidden defects	Impact-echo; Infrared thermography (ASTM D4788); impulse-response; radiography; GPR	Sounding (ASTM D4580); pulse-echo; SASW; intrusive drilling and borescope	Assessment of reduced structural properties; extent and location of internal damage and defects; sounding limited to shallow delamination
Concrete component thickness	Impact-echo (I-E); GPR (ASTM D4748)	Intrusive probing	Verify thickness of concrete; provide more certainty in structural capacity calculations; I-E requires knowledge of wave speed, and GPR of dielectric constant
Steel area reduction	Ultrasonic thickness gage (requires direct contact with steel)	Intrusive probing; radiography	Observe and measure rust and area reduction in steel; observe corrosion of embedded post-tensioning components; verify location and extent of deterioration; provide more certainty in structural capacity calculations

The degree of confidence of the core test depends on the number of tests, which must be minimal. The relation between number of cores and confidence is shown in Table 9.2.

Before you choose the location of the sample, first of all you must define the location of the steel bars to assist you select a sample away from the steel bars to avoid the possibility of taking a sample containing steel reinforcement bars.

We must carefully determine the places to preserve the integrity of structure and therefore this test should be performed by an experienced engineer. To conduct such

TABLE 9.2
Number of Cores and Deviation in Strength

Number of Cores	Deviation Limit between Expected Strength and Actual Strength (Confidence Level 95%) (%)
1	+12
2	+6
3	+4
4	+3

Inspection Methods

an experiment, one should be taking precautions, determining the responsibility of individuals, and reviewing the non-destructive testing which had been conducted in an accurate manner.

Figures 9.1 and 9.2 present the process of taking the core from reinforced concrete bridge girders.

Core Size

Note that the permitted diameter is 100 mm in the case of maximum aggregate size 25 mm, and 150 mm in case the maximum aggregate size does not exceed 40 mm. It is preferably using 150 mm diameter whenever possible as it gives more accurate results, as shown in the Table 9.3, which represents the relationship between the dimensions of the sample and potential problems. So this table should be consulted to choose the reasonable core size.

FIGURE 9.1 How to take a core.

FIGURE 9.2 Shape of core sample.

TABLE 9.3
Core Size with Possible Problem

Test	Diameter (mm)	Length (mm)	Possible Problem
One	150	150	May contain steel reinforcement
Two	150	300	May cause more cutting depth to concrete member
Third	100	100	Not allow if the maximum aggregate size 25 mm May cut with depth less than required
Last	100	200	Less accurate data

Some researchers stated that the core test can be performed with a core diameter of 50 mm in the case of maximum aggregate size of not more than 20 mm overall. It was noted that small core sizes give different results than large sizes.

Because of the seriousness of the test and not to allow taking high number of samples, so it needs well supervision when taking the sample. Moreover, the laboratory test must be certified and the test equipment must be calibrated with a certificate of calibration from certified company.

Sample extraction are conducted using pieces of a cylinder which is a different country-cylinder equipped with ransom of special alloy mixture with diamond powder to feature pieces in the concrete during the rotation of the cylinder through the body. The precautions taken should be a proper match for the sampling method and consistent pressure should be applied carefully depending on the expertise of the technician.

After that, the core will be filled with dry concrete of suitable strength or with grouting, which is a popular method. Other solution depends on using epoxy and injecting it into the hole and then inserting a concrete core of the same size to close the hole.

In any case, the filling must be done soon after making the cutting, so the material of filling will be standing by next to the technician who did the cutting, as this core affects the integrity of the structure.

Every core must be examined and photographed in the lab, and noted gaps must be identified within the core as a small void if measured between 0.5 and 3 mm, void if the average measured between 3 and 6 mm, or a big void if measured more than 6 mm. This should also be examined whether the core of nesting and determine the shape, kind, and color gradient of aggregates any apparent qualities of the phenomenon of sand as well.

In the laboratory, the dimensions, weight of each core, the density, steel bar diameter, and distance between the bars will measure.

Sample Preparation for Test

After cutting the core from the concrete element, process the sample for testing by leveling the surface of the core, and then take the core that has a length not less than 95% of the diameter and not more than double the diameter.

Figure 9.2 shows the shape of the core sample after cutting from the concrete structure member directly.

For leveling the surface, use a chainsaw spare concrete or steel cutting disk. After that, prepare the two ends of the sample by covering them with mortar or sulfide,

Inspection Methods

and submerge the sample in the water at a temperature of 20 ± 2 °C for at least 48 h before testing the sample.

The sample is put in a machine and a suitable load is applied with the rate of regular and continuous range of 0.2–0.4 N/mm² until it reaches the maximum load at which the sample has been crushed.

Determine the estimated actual strength for cube by knowing the crushing stress that is obtained from the test using the following equation, where λ is the divided the core length to its diameter.

- In the case of horizontal core, the strength calculation will be as follows:

$$\text{Estimated actual strength for cube} = \frac{2.5}{1/\lambda} + 1.5 \times \text{core strength}$$

where λ is the core length/core diameter.

- In the case of vertical core, the strength calculation will be as follows:

$$\text{Estimated actual strength for cube} = \frac{2.3}{1/\lambda} + 1.5 \times \text{core strength} \tag{9.1}$$

In the case of existing steel in the core perpendicular to the core axis, the previous equations will be multiplied by the following correction factor:

$$\text{Correction factor} = 1 + 1.5 \frac{s\varphi}{LD} \tag{9.2}$$

where
L is the core length,
D is the core diameter,
s is the distance from steel bar to edge of core, and
φ is the steel bar diameter.

Cores are preferred to be free of steel, but if steel is found, you must use the correction factor considering that it is taken only in the event that the value ranges from 10% to 5%. We must agree to use the results to core, but if the correction factor is more than 10%, the results of the cores cannot be trusted and then you should take another core.

The cores samples are often taken after 28 days, so it is in mind that it must provide higher strength than that the data of the cylinder test after 28 days but practically speaking the situation is different as shown on the following Table 9.4 for study performed by Yuan et al. in 1991 for reevaluation of cores strength for high strength concrete by comparing between the standard cylinders and cores taken from column cured using a sealing compound. It can be seen that in situ concrete often gains little strength after 28 days (Table 9.4).

When examining the test results, the following points must be taken into account when evaluating the results:

- Before the test, submerge the sample in water, leading to a decrease in strength to up to about 15% for dry concrete.
- The equation to calculate the expected actual concrete strength does not take into account any differences in direction between the core and the standard cube direction.

TABLE 9.4
Relation between Standard Cylinder and Cores

Age (days)	Standard Cylinder Strength (MPa)	Cores Strength (MPa)	Fc (Core)/fc (Cylinder at 28)
7	66	57.9	0.72
28	80.4	58.5	0.73
56	86.0	61.2	0.76
180	97.9	70.6	0.88
365	101.3	75.4	0.94

- It states that the concrete is acceptable if the average strength to the cores is at least 75% of the required strength and the calculated strength for any core is less than 65% of the required strength.

It states that in the case of prestressed concrete, the concrete strength is acceptable if the average strength to the cores is at least 80% of the required strength and the calculated strength for any core is less than 75% of the required strength.

9.1.2 REBOUND HAMMER

This is a non-destructive test for determining the estimate concrete compressive strength.

This is the most common method as it is easy to do and so cheap compared with other tests, but gives less precise outcomes (see Figure 9.3 for an example of a rebound hammer).

This test relies on measuring the concrete strength by measuring the hardening from the surface. It will be used to identify the concrete compressive strength of the concrete member by using calibration curves of the relationship between reading the concrete hardening and concrete compressive strength. Most commonly, these give impact energy of 2.2 N/mm. There is more than a way to show results based on the manufacturer. In some cases, the reading will be analog or digital number, or by connecting to a memory to record the readings.

Figure 9.4 shows the rebound hammer with an indicator number (also known as the rebound number) that measures concrete strength based on the element location.

Inspect the device before using it through the calibration tools that are attached with the device. The calibration should be within the allowable limit based on manufacturer's recommendation.

The first and most important step in the test is cleaning and smoothing the concrete surface at the sites that will be tested by measuring an area of about 300 × 300 mm and, preferably, test on the surface that has no change after casting or the surface that had not any smoothing during casting process.

On the surface to be tested, draw a net of perpendicular lines in both directions by 2–5 cm apart, and the intersection points will be the points to make the test on it and the test point in any case must be away from the edge by about 2 cm.

Inspection Methods

FIGURE 9.3 Rebound hammer.

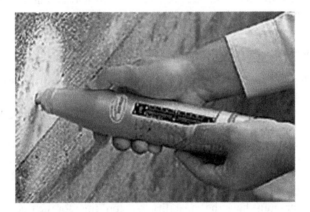

FIGURE 9.4 Test and read the number.

The surface must be cleaned before performing the test and the rebound hammer is perpendicular to the surface.

As a result, the following recommendations are made during the test:

- The hammer must be perpendicular to the surface that will be tested at any conditions because the direction of the hammer affects the value of rebound number as a result of the impact of hammer weight.
- Note that the wet surface gives a reading of rebound hammer less than reading a dry surface significantly by up to 20%.
- The tested concrete member must be fixed and does not vibrate.
- You must not use the curves for the relationship between concrete compressive strength and rebound number as given from manufacturer directly, but you must calibrate the hammer by taking the reading on concrete cubes and crushing the concrete cubes to obtain the calibration of the curves. This calibration is important to be done from time to time as the spring inside the rebound hammer lost some of its stiffness with time.

- You must use one hammer only when you making a comparison between the quality of concrete at different sites.
- Type of cement affects the readings, as in the case of concrete with high-alumina cement, which can yield higher results than those of concrete with ordinary Portland cement by about 100%.
- Concrete with sulfate-resistant cement can yield results in about 50% less than those of concrete with using ordinary Portland cement.
- Higher cement content gives lower reading than concrete with less cement content; in any case it will give the gross error of only 10%.

9.1.2.1 Data Analysis

The number of reading must by high enough to give reasonable accurate results. The minimum number of reading is 10 but usually we take 15 reading.

The extreme values are excluded and take the average for the other remaining values. From this, the concrete compressive strength will be known and compared the results with the required concrete strength.

9.1.3 Ultrasonic Pulse Velocity

This test is one of the non-destructive testing, as its concept is to measure the speed of transmission of ultrasonic pulses through the construction member by measuring the time required for the transmission of impulses and by knowing the distance between the sender and receiver the pulse velocity can be calculated.

The calibration of these velocities is done by knowing the concrete strength and its mechanical characteristics. Then, use it for any other concrete with the same procedure of identifying compressive strength, the dynamic and static modulus of elasticity, and the Poisson ratio.

The equipment must have the capability to record time for the tracks with lengths ranging from 100 to 3000 mm accurately +1%.

The manufacturer should define the way of working of the equipment in different temperature and humidity requirements. The device must have a power transformer sender and receiver of natural frequency vibrations between 20 and 150 kHz, bearing in mind that the frequency appropriate for most practical applications in the field of concrete is 50–60 kHz.

There are different ways for wave transmission, for example, surface transmission, as shown in Figures 9.5–9.7; semi-direct transmission, as shown in Figure 9.6; and wave transmission, as shown in Figure 9.7.

The UT equipment is composed of two rods of metal with lengths of 250 and 1000. The first is used in the determination of zero of the measurement and the second is used in the calibration. In both cases, the time of the passage of waves through each rod known is obtained.

Hence, connect the ends of the rods by sender and receiver in an appropriate way and measure the time for pulse transmission and compare it with the known reading; the smaller road, if there are any deviations, adjusts the zero of the equipment to provide the known reading.

Inspection Methods

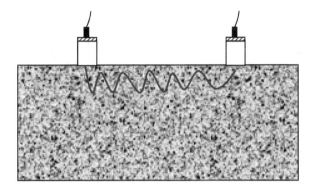

FIGURE 9.5 Surface transmission.

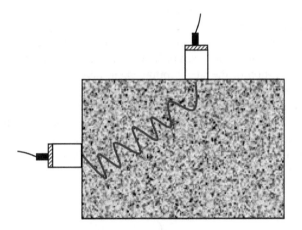

FIGURE 9.6 Semi-direct transmission.

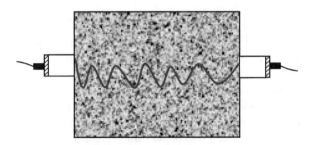

FIGURE 9.7 Direct transmission.

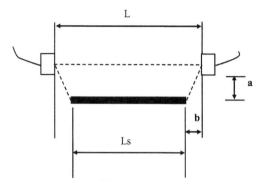

FIGURE 9.8 UT wave parallel to the steel bars.

For the long bar, the same way is used to define the result accuracy. In this case the difference between the two readings should not be more than ±0.5% for the measurements to be within the required accuracy.

It is worth to mention that the wave transmission velocity in steel is twice than that in concrete, so in the case of steel bars in concrete member, testing will influence the accuracy of the reading as it will be high value for the wave impulse velocity. To avoid that, the location of the steel reinforcement must be defined previously with respect to the path of the ultrasonic pulse.

The correction of the reading, in case of steel bars parallel to the path of the pulse wave as shown in the Figure 9.8 must considered. The calculation of the pulse velocity will be as shown in the following equation. By knowing the wave path, the wave velocity in steel, and the distance between the two ends, time of transmission will be as follows:

$$V_C = K V_n \tag{9.3}$$

$$K = \gamma + 2\left(\frac{a}{L}\right)(1-\gamma^2)^{0.5} \tag{9.4}$$

$$L_s = L - 2b \tag{9.5}$$

V_m is the pulse velocity obtained from the transmission time from the equipment,
V_c is the pulse velocity in concrete, and
γ is a factor whose value varies according to steel bar diameter.

The effect of the steel bar on the reading can be ignored if the diameter is equal to 6 mm or less or in case the distance is far away between the steel bar and end of the equipment.

If the steel reinforcement bars axis is perpendicular to the direction of pulse transmission, as shown in Figure 9.9, the effect of the steel will be less in this case on the reading. The effect can be considered zero if we use transmission source of 54 KHZ and the steel bar has a diameter less than 20 mm. The above equation is used if the frequency is less and the diameter is higher than 20 mm and by change the value of γ according to the bar diameter by the data which are delivered by the equipment.

Inspection Methods

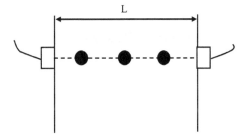

FIGURE 9.9 Wave of UT perpendicular to the steel bars.

There are other factors that influence the measurement such as the temperature, thawing, and concrete humidity, and these effects must be considered (Table 9.5).

Most Common Error
1. Ignore using the reference bar to adjust the zero and this will impact the accuracy of the results.
2. The concrete surface that is well-leveled and smoothed after pouring may have properties different from the concrete in the core of the member and therefore avoid it as much as possible in the measurements. In the absence of the possibility of avoiding, one must take into account the impact of the surface.
3. Temperature affects the transmission ultrasonic velocity according to the previous table, which must be taken into account with increases or decreases in temperature over 30°C.
4. When comparing the quality of concrete of various components of the same structure, similar circumstances should be taken into account in all cases in terms of the composition of concrete, moisture content, and age, temperature, and type of equipment used. There is a relationship between the quality of concrete and the speed of the pulse, as shown in Table 9.6.

The static and dynamic moduli of elasticity can be defined by knowing the transmission pulse velocity in the concrete, as shown in Table 9.7.

TABLE 9.5
Temperature Effect on Pulse Transmission Velocity

Temperature (°C)	Percentage Correction of the Velocity Reading (%)	
60	+5	+4
40	+2	+1.7
20	0	0
0	−0.5	−1
4	−1.5	−7.5

TABLE 9.6
Relation between Concrete Quality and Pulse Velocity

Pulse Velocity (km/s)	Concrete Quality Degree
>4.5	Excellent
4.5–3.5	Good
3.5–3.0	Fair
3.0–2.0	Poor
<2.0	Very poor

TABLE 9.7
Relation between Elastic Modulus and Pulse Velocity

| | Elastic Modulus (MN/mm²) | |
Transmission Pulse Velocity (km/s)	Dynamic	Static
3.6	24,000	13,000
3.8	26,000	15,000
4.0	29,000	18,000
4.2	32,000	22,000
4.4	36,000	27,000
4.6	42,000	34,000
4.8	49,000	43,000
5.0	58,000	52,000

9.1.4 LOAD TEST FOR CONCRETE MEMBERS

This test is performed in the following conditions:

- If the core test gives results of concrete compressive strength lower than those of characteristic concrete strength, which is defined in design.
- If this test is included in the project specifications
- If there is a doubt in the ability of the concrete structure member to withstand design loads.

This test is usually performed on the slabs and, in some cases, at the beams. The summary of this test is to expose the concrete slab to a certain load and then remove the load. During this period, it measures the deformation on the concrete member as deflection or presence of cracks and compares it with the allowable limit in the specifications.

9.1.4.1 I-Test Procedure

This is done by loading the concrete member with a load equal to the following:

$$\text{Load} = 0.85(1.4 \text{ dead load} + 1.6 \text{ live load}) \tag{9.6}$$

Inspection Methods

The applying load will be done by using sacs of sand or concrete blocks. In the case of sand, sacs are calibrated to at least 10 sacs for every span, about 15 m² through direct weight of the sacs. These sacs are chosen randomly to determine the weight of an average sac. Then put these sacs on the concrete member that will be tested and take into consideration to be a distance between vertical sacs to prevent arch effect.

As for the concrete blocks, their weight should be measured and they should be calibrated. Also the horizontal distance between them should be taken into account to avoid influencing the arching effect.

It is important to identify of the adjacent elements that have an impact on the structure element to be loaded also in order to obtain the maximum possible deformation for the test member.

Before load processing, the location of the test must be defined by identifying the places where the gauges will be placed, as well as calculating the actual dead load on the concrete member, through the identification weight of the same member. In addition, coverage such as tiles, which will be installed on a slab of concrete, as well as lower coverage should include, for example, plastering or weight of any kind of finishing work.

The location of the measurement unit is shown as an example for slab test in Figure 9.10. The figure illustrates the following specifications:

1. It places in the middle of the span; placed beside it is another as a reserve, as shown in the figure.
2. Put another measurement device at a quarter the span from the support; and the consultant engineer must define the other reasonable location for measurement.
3. The measurement devices must be calibrated and certified before use, preferably less sensitivity of 0.01 mm and its scale is about 50 mm.
4. We must have devices that measure the cracks width and this device must have an accuracy of 0.01 mm.

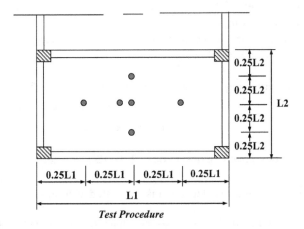

FIGURE 9.10 Location of measurement devices.

Test Procedure
- Define load test = 0.85 (1.4 dead load + 1.6 live load)—dead load that has already affected the member.
- Take the reading of the deflection before starting the test (R1).
- Start to put 25% of the test load and avoid the arching effect or any impact load.
- Read the measurement for the effect of 25% of the load and visually inspect the member to see if there are any cracks. If cracks are present, measure its width.
- Repeat this procedure three times; for each time, increase 25% of the load.
- Record the time for putting the last load and the last deflection reading and crack thickness.
- After 24 h of load affects the time record, draw the location of the cracks, the maximum crack thickness, and the deflection reading (R2). Then remove the load gradually and avoid any impact load.
- After removing all the load, measure the deflection reading and crack width.
- After 24 h from removing the load, record the measurement and the reading is (R3) and record the crack width.

Result Calculations

1. The maximum deflection after 24 h from the load effect is as follows:

$$\text{Maximum deflection} = \frac{(\text{first measurement after passing 24 h from load effect-reading before load effect})}{\text{device sensitivity}} \quad (9.7)$$

$$\text{Maximum deflection} = (R2 - R1) \times \text{device sensitivity} \quad (9.8)$$

When any problem occurs in the use of first device, the second device reading should be used. If the reading of the two devices is close, take the average of the two readings.

2. The remaining maximum deflection after 24 h of completely removing the load will be from the following equation:

$$\text{Maximum remaining deflection} = (\text{reading of the device after 24 h from removing the load} - \text{reading before load effect}) \times \text{device sensitivity} \quad (9.9)$$

$$\text{Maximum remaining deflection} = (R3 - R2) \times \text{device sensitivity} \quad (9.10)$$

3. The maximum recovery deflection is calculated as follows:

$$\text{The maximum deflection recovery} = \text{max. deflection} - \text{the max. remaining deflection} \quad (9.11)$$

Inspection Methods

The maximum deflection after 24 h from loading and the recovery maximum deflection is as shown in the previous figure.
4. The relation between the load and maximum deflection in the case of loading and uploading is drawn.
5. The maximum crack thickness is calculated after 24 h from load effect and 24 h after removing the load.

Acceptance and Refusing Limits
- Calculate the maximum allowable deflection for the member as follows:

$$\text{The maximum allowable deflection} = \frac{L^2}{2t}, \text{cm} \qquad (9.12)$$

where L is the span of the member in meter. The shorter span in the case of flat slab and short direction for solid slab and in the case of cantilever it will be twice the distance from the end of cantilever and the support face; t is the thickness of the concrete member in cm.

Compare between the maximum deflection recorded after 24 h from load effect and the allowable maximum deflection. Thus, we have three outcomes:

- If the maximum deflection after 24 h from load effect is less than the allowable maximum deflection from the previous equation, then the test is successful and the member can carry load safely.
- If the maximum deflection after 24 h from load effect is higher than the allowable maximum deflection, then the recover deflection after 24 h from removing load must be equal or higher than 75% of maximum deflection.

$$\text{Recovery deflection} \geq 0.75 \text{ maximum deflection}$$

If this condition is verified, then the member is considered to be successful.

- If the recovery is less than 75% of the maximum deflection, the test should be repeated by the same procedure, but after 72 h from removing the load from the first test.
- After repeating the test for second time using the same procedure and precautions, this concrete structure member will be refused if not verified by using the following two conditions:
  1. if the recovery deflection in the second test is less than 75% of the maximum deflection after 24 h from load effect in second test
  2. If the recorded maximum crack thickness is not allowed

9.1.5 Comparison between Different Tests

From the above discussion, the different methods to determine the hardened concrete strength have advantages and disadvantages, as summarized in Table 9.8.

TABLE 9.8
Comparison between Different Test Method

Test Method	Probable Damage	Precaution Requirement
Overload test	Possible member loss	Member must be isolated or allowance of distributed the load to the adjacent members
		Extensive safety precaution just in the case collapse failure
Cores	Holes to be made good	Limitations of core size and numbers
		Safety precaution for critical member
Ultrasonic	Non	Need two smooth surfaces
Rebound hammer	Non	Need a smooth surface

TABLE 9.9
Performance Comparison between Different Methods

Test Method	Damage to Concrete	Representative to Concrete	Accuracy	Speed of the Test	Cost
Overload test	Variable	Good	Good	Slow	High
Cores	Moderate	Moderate	Good	Slow	High
Ultrasonic	Non	Good	Moderate	Fast	Low
Rebound hammer	Unlikely	Surface only	Poor	Fast	Very low

- Moreover, these tests are different from their costs, their representation of the concrete member, and accuracy of the measured strength to the actual concrete strength; this comparison is illustrated in Table 9.9.
- After obtaining the result of the tests, one can find that the data of strength are lower than the concrete strength specified in the drawings or project specifications. The value of concrete strength in the drawing is the standard cube (cylinder) compressive strength after 28 days. Note that the cube or the cylinder will be poured and compacted based on the standard; the curing process is for 28 days so it is more different that will happen in the site. For example, the curing will not be for 28 days and the concrete slabs or columns cannot be immersed in water continuously for 28 days. The temperature on-site is surely different and varied than that in the laboratory. In addition, the concrete pouring and compaction method is different for the cubes or cylinder than that in the actual member size. All these variations were taken into consideration in the factors in the design codes, but the test measures the strength after complete hardening to the concrete.

9.1.6 DEFINE CHLORIDE CONTENT IN HARDEN CONCRETE

The determination of the maximum chloride content in concrete is an important factor in maintaining the reinforced concrete structure from the corrosion of the steel

Inspection Methods

bars. Table 9.10 shows the limits of standardization of the content of ions dissolved chlorides in the concrete.

The American code makes clear the limits of different contents in chlorides and there are limits to the concrete content of ions dissolved chlorides as in the ACI 318R-89, and chloride ion limits as stated in ACI Committee 201,357,222 result of the components of concrete or as a result of the content of chlorides total result of chlorides penetrate reinforced concrete in the life of structure. American specifications ACI COMMITTEE357 recommended that water used in the concrete mixture should not contain more than 0.07% chloride in the case of reinforced concrete or 0.04% in the case of pre-stress concrete.

European Code in the year 1992 (ENV206) stated the limits of chlorides in concrete and identified these limits according to the type of application, and these limits are 0.01% of the weight of cement for plain concrete, 0.04% of the weight of cement for the reinforced concrete, and 0.02% of the weight of cement concrete for pre-stress concrete.

The EU specifications disallow the use of any additives that have chloride or calcium chloride in reinforced concrete or prestress concrete (Table 9.11).

TABLE 9.10
Maximum Allowable Solved Chloride Ions

Condition around Concrete	Maximum Dissolved Chloride Ions in Concrete Water – % from Cement Weight
Reinforced concrete exposed to chloride	0.15
Dry reinforced concrete and totally protected from humidity during use	1.0
Different structure members	0.3

TABLE 9.11
ACI Recommendation for the Maximum Acceptable Chloride Ions

Type of Member	Dissolved[a] (in Water)	Total[b]	Dissolved[c] (in Acid)	Total[d]
Prestress concrete	0.06		0.06	0.08
Reinforced concrete exposed to chloride during use	0.15	0.1	0.1	0.2
Dry reinforced concrete	1.0			
Different structure members	0.3	0.15		

[a] (ACI 318R-89) ACI Building Code.
[b] ACI Committte202.
[c] ACI Committee 357.
[d] ACI Committee 222.

9.1.7 Concrete Cover Measurements

The thickness of the concrete cover is measured in modern construction so as to ensure that the thickness of the concrete cover conforms with the specifications.

The process of measuring the thickness of the concrete cover in structures was begun when corrosion, due to less cover thickness, caused an increase in the corrosion rate. This resulted from chlorides or carbonation where it propagated inside the concrete causing the speed of steel corrosion. Also, the lack of cover thickness helps the propagation of moisture and oxygen, which are the main basics for the corrosion process.

The measurement of the concrete cover thickness explains the causes of corrosion and identifies areas that have the capability to corrode faster.

It is noted that the measurement of the thickness of the concrete cover needs to be defined as axis y, x in order to determine the thickness of the concrete cover at every point on the structure.

The equipment that measures the thickness of the concrete cover is simple and high tech and one can obtain measurement readings in the form of numbers. In the bridges inspection, it can be used radiograph but it is high-cost (Cadry Gamnon, 1992; Bungey, 1993).

The magnetic cover method is a simple method, but can be affected by the distance between steel bars, and the thickness of the concrete cover can influence the readings significantly. Because the reason for that, this method depends on supply electricity through 9-volt battery, while the second side of measurement for the potential voltage envelope when completed by the buried steel bars.

Figure 9.11 illustrates the magnetic cover test device on the concrete surface and Figures 9.12 and 9.13 present the direction of the device from which we can obtain the maximum and minimum signal.

The British standard is the only standard that cares about measurement of concrete cover after construction in BS 1881, Part 204. In 1993, Alldred studied the accuracy of the measurement of the cover when there were more steel bars close to each other, and he suggested using more than one head of measurement as that would increase

FIGURE 9.11 Concrete cover measurement.

Inspection Methods

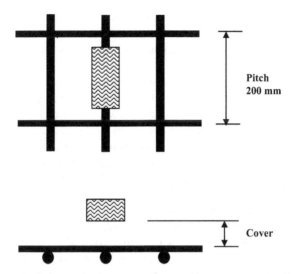

FIGURE 9.12 Device location on concrete surfaces.

FIGURE 9.13 The proper device location.

the accuracy of the reading and the small heads have impact on the accuracy of the equipment.

Therefore, the problem with this method is that dense steel reinforcement in the concrete section will give inaccurate data. The equipment will be calibrated based on the existing steel bars as the reading can be affected by the type of steel. The person who works with this equipment must be competent and aware for anything that can affect the reading, such as bolts, steel wires, etc.

9.1.8 Measurement of Carbonation Depth

Carbonation depth is formulated in the propagation of carbon dioxide inside the cover of the concrete which changes the concrete alkalinity value and reduces its value. The concrete alkalinity value is expressed by pH values equal to 12–13, with carbon dioxide propagation and causing reaction which reduces the alkalinity and this will affect the steel surface and then damage the passive protection layer around the steel bars and then the corrosion process will start.

Therefore, it is very important to define the depth of the concrete that transforms to carbonation and how it is far away or closest from the steel bars.

This test is performed by spraying the surface of currently broken concrete or let it break by special tools to obtain the carbonation depth by Phenolphthalein dissolved in alcohol. This solution color becomes pink when touching the surface of concrete with alkalinity value of pH about 12–13.5, and the color turns into a gray or blue if concrete loose its alkalinity and pH value is less than 9. In this case, steel bars lose their passive protection layer.

The accuracy of the test depends to be sure that the part which will test newly broken and must be done during the test to test whether beams, columns or slabs. After measuring the carbonation depth in the concrete cover and the distance from carbonation depth to the steel bar, if it reaches to the steel bars or not so, it is easy to evaluate the corrosion risk the steel reinforcement as shown in Figure 9.14.

The best solution is to use Phenolphthalein with alcohol and water to be 1 mg Phenolphthalein with 100 mm alcohol/water (50:50 mixing ratio) or alcohol more than water (Building Research Center in 1981; Parrott, 1987) if the concrete is completely dry humidifying the surface by water.

If the thickness of carbonation about 10–5 mm, it breaks the passive layer to the steel bar from 5 mm from the presence of change color of the surface of concrete.

It is important to note that some of the aggregates or some concrete mixtures have a dark color which makes Phenolphthalein reading very bad, and, therefore, we must take into account the surface clean when testing.

This test must be done in the areas accessible to the work of the breakers required in the concrete, as well as facilitate the work of repair easily for the part, which was broken.

9.1.9 CHLORIDES TEST

Chlorides tests (AASHTO T260-84, 1984; ASTM D1411-82, 1982) rely on an analysis of the samples of the concrete powder to determine the quantity of chloride. This test is done by making a hole with drilling machine to rotary dig inside the concrete and extracting concrete powder output of the hole or through the broken part of the concrete.

It must be taken several separate samples at different depths. The depth of the hole will vary to increase the accuracy and always taken from 2 to 5 mm. The first

FIGURE 9.14 Carbonation depth measurement.

Inspection Methods

5 mm from the surface of concrete is where the chlorides concentration often is very high, particularly in structures exposed to sea water and chlorides

The holes are made by special devices to collect concrete powder product of the process of drilling the hole. The concrete powder which will be taken at every drilling depth will be added to a solution of acid, which is determined by the amount of chlorides concentration. There are two principal ways of measuring in situ (Quanta Strips) and the method to determine the electron-ion (Specific Ion Electron) and the recent experience more accurate, but we need high-cost equipment.

In the experiments of chlorides concentration, professionally trained workers are needed.

When the readings of chloride concentration at various depths have been assembled, draw the shape to chloride concentration with depth from the surface of concrete into the concrete.

From the figure, one could determine whether chlorides are within the concrete or from the impact of air and environmental factors affecting the concrete from abroad, as shown in Figure 9.15.

It is shown from Figure 9.15 that in the case of the affect of chlorides from outside that, in the beginning the chlorides concentration is high and then gradually decrease with the depth until reach constant value.

On the other hand, if the chloride exists inside the concrete, such as during concrete mixing due to use of salt water or an aggregate with high chloride the figure shows that near the surface the chloride concentration is minimal and increases with depth until it reaches constant values with the depth.

The chlorides test method is done through melted in acid and calibration work after that, based on specifications (BS 1881, Part 124). In addition to the method of the melting, the melting acid can work in the water; these tests follow ASTM D1411, 1982 and AASHTOT260, 1984.

The effect of chloride concentration on the steel reinforcement will be defined if the chloride concentration is exceeded 0.6 from hydroxyl concentration that causes failure in the passive layer around the steel reinforcement. This ratio is approximately

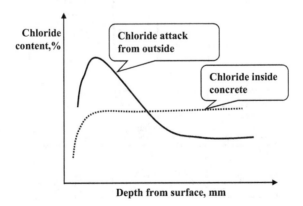

FIGURE 9.15 Comparison between chloride attack from outside and with the concrete.

0.2%–0.4% chloride by weight of cement, 1 Ib/Yard cubic concrete, or 0.5% chloride by the weight of concrete.

The figure for the distribution of chloride concentration with depth can determine the degree of impact of chlorides in the steel reinforcement and the impact of chlorides if it is from inside the concrete itself or from outside.

9.1.9.1 Half Cell

This measurement unit is traditionally used in oil and gas plants to monitor the cathodic protection system for the pipeline and the facilities, and you will find that a well-trained corrosion team can assist you in these measurements.

This system is used to determine the bars that have lost its passive protection. the steel bars in concrete that have lost the passive protection due to exposed to carbonating or some existing chlorides, which produces a chemical that reacts with the alkaline around the steel, which loses its passive layer.

The half-cell equipment (ASTM C876, 1991) is composed of a rod of metal in solution by the same metal ions such as a copper rod in saturated solution of copper sulfate. This is considered to be one of the most common types of equipment and there are some other pieces of equipment, such as bars of silver in silver chloride solution, can be linked through the voltmeter and other part will be connected to the steel bars as in the following figure (Figure 9.16).

If the steel bars have a passive protection layer, the potential volt is small, with a range from 0 to 200 MV copper/copper sulfate. If there has been a breakdown of the passive protection layer and some quantity of steel has melted as a result of movement of ions, the potential volt is about 350 MV; when the value is higher than 350 MV, it means that the steel has already started to corrode.

It is recommended to use half-cell equipment which consists of silver and silver chloride solution or mercury and mercury oxide solution. The equipment which

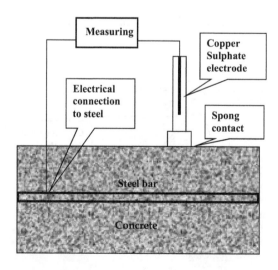

FIGURE 9.16 Half cell machine.

Inspection Methods

consists of copper and copper sulfate solution is used, but its usage is not recommend because it requires regular maintenance and thus readings contain some mistakes.

To do a survey for large concrete surface area determine the bars that have been corroded. As a result is a contour line on the concrete surface and every line present an equal value of potential volt.

The measurement of a half-cell to the steel reinforcement bars is to define the probability of causing corrosion to the steel bars, and, in the ASTM C867 specification, present a method to clarify the data, which obtained from measuring the potential volts. These values are shown in Table 9.12 for copper/copper sulfate equipment and also silver/chloride silver equipment and explain the condition of corrosion in each case of corresponding value.

The inaccuracy of such measurements is a result of the presence of water, which increases the negative values without happened corrosion in steel as wetting concrete columns or walls give high negative values for potential volt regardless of the degree of corrosion steel and negative voltage difference significantly when underwater in the offshore structures in spite of the lack of oxygen, which reduces the rate of corrosion (Figures 9.17 and 9.18).

TABLE 9.12
ASTM Specification for Steel Corrosion for Different Half Cell

Silver/Silver Chloride (mV)	Copper/Copper Sulfate (mV)	Standard Hydrogen Electrode (mV)	Calomel (mV)	Corrosion Condition
>−106	>−200	>+116	>−126	Low (10% risk of corrosion)
−106 to −256	−200 to −350	+116 to −34	−126 to −276	Intermediate corrosion risk
<−256	<−350	<−34	<−276	High (<90% risk of corrosion)
<−406	<−500	<−184	<−426	Severe corrosion

FIGURE 9.17 Half cell test.

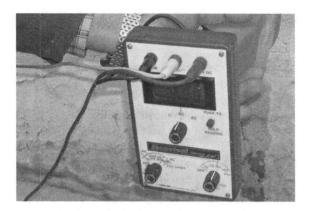

FIGURE 9.18 Half cell reading.

9.2 STEEL STRUCTURE

9.2.1 Introduction

Non-destructive testing is very essential technique from the steel industry. These methods can guarantee that the work is as per to the standard and discussed in Chapter 7. The main activity that needs a special focus is the welding, so it needs to define the reasonable technique that can be used based on the standard to guarantee that the product is as per to the standard and project specification. The men who perform these tests should be certified and based on ASNT association there is a three level of certificate.

9.2.2 Radiographic Test

The most important technique in the NDT is a radioactive test because is it more accurate and documented, but it is more harmful to human if safety precautions are not closely followed. The basis of this test is to use the short-wavelength electromagnetic radiation with high-energy photons to penetrate through the materials.

It is worth to mention that this test is used in new construction only and it is rarely to use in evaluating the existing structure. But in the case of failure you can return to the project file and define the welding radiographic.

9.2.3 General Welding Discontinuities

To know the discontinuity by seeing the film, this is need a well training to the inspector and the following is types of discontinuities and the correspond film image as a reference.

In case of cold lap which is a condition where the weld filler metal does not properly fuse with the base metal or the previous weld pass material. . This happened due to the arc does not melt the base metal sufficiently and causes the slightly molten puddle to flow into the base material without bonding.

Inspection Methods 237

Figure 9.19 presents the cold lap with sketch and radiographic film.

As a reason of gas entrapment in the solidifying metal. Porosity can take many shapes on a radiograph but often appears as dark round or irregular spots or specks appearing singularly as shown in Figure 9.20. Sometimes, porosity is elongated and may appear to have a tail. This is the result of gas attempting to escape while the metal is still in a liquid state and is called wormhole porosity. All porosity is a void in the material and it will have a higher radiographic density than the surrounding area.

Figure 9.21 presents a cluster porosity which is caused when fluxcoated electrodes are contaminated with moisture. The moisture turns into gas when heated

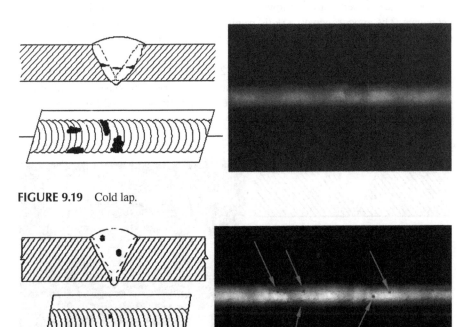

FIGURE 9.19 Cold lap.

FIGURE 9.20 Porosity.

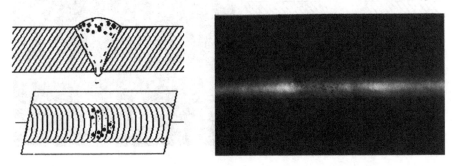

FIGURE 9.21 Cluster porosity.

and becomes trapped in the weld during the welding process. The shape of cluster porosity appears the same as the regular porosity in the radiograph but the indications will be grouped close together.

Slag inclusions are nonmetallic solid material entrapped in weld metal or between weld and base metal. In a radiograph, dark, jagged asymmetrical shapes within the weld or along the weld joint areas are indicative of slag inclusions as shown in Figure 9.22.

As discussed in Chapter 7, the incomplete penetration occurs when the weld metal fails to penetrate the joint. It is one of the most objectionable weld discontinuities. Lack of penetration allows a natural stress riser from which a crack may propagate. The appearance on a radiograph is a dark area with well-defined, straight edge that follows the land or root face down the center of the weldment, as shown in Figure 9.23.

Incomplete fusion is shown in Figure 9.24 where the weld filler metal does not properly fuse with the base metal. Appearance on radiograph is usually a dark line or lines oriented in the direction of the weld seam along the weld preparation or joining area.

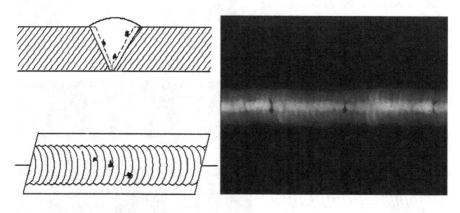

FIGURE 9.22 Slag inclusions.

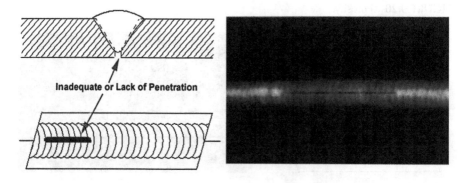

FIGURE 9.23 Incomplete penetration.

Inspection Methods 239

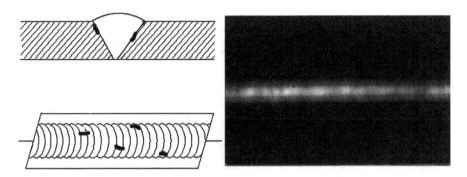

FIGURE 9.24 Incomplete fusion.

In the case of internal concavity or suck back, it is happened when the weld metal has contracted as it cools and has been drawn up into the root of the weld. On a radiograph, it looks similar to a lack of penetration but the line has irregular edges and it is often quite wide in the center of the weld image (Figure 9.25).

The undercut near to the crown or the root is shown in Figure 9.26 for A, and B cases, respectively.

The offset or mismatch between the two plates that is welded is shown in Figure 9.27. The radiographic image shows a noticeable difference in density between the two pieces. The difference in density is caused by the difference in material thickness. The dark straight line is caused by the failure of the weld metal to fuse with the land area.

The inadequate weld reinforcement or excess weld reinforcement is shown in Figure 9.28a, b, where the thickness of weld metal deposited is less than the thickness of the base material or more than that mentioned in the engineering drawings.

9.2.4 Ultrasonic Test

This is a very effective technique to inspect the existing structure as you can measure the steel thickness after corrosion and define if there are any cracks.

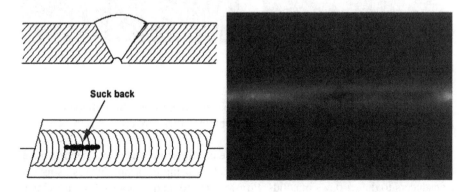

FIGURE 9.25 Internal concavity.

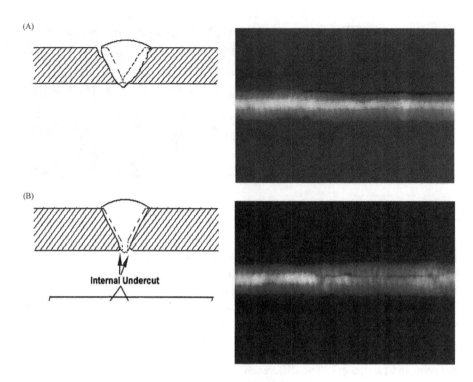

FIGURE 9.26 Under cut for crown (A) and the root (B).

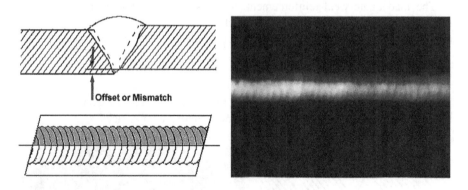

FIGURE 9.27 Offset between two plates.

The main principal of this technique is to send sound wave through the materials and receive the sound wave. Thus, we can obtain if there are cracks in the steel and the steel thickness.

In 1929 and 1935, Sokolov studied the use of ultrasonic waves in detecting metal objects. Mulhauser, in 1931, obtained a patent for using ultrasonic waves using two transducers to detect flaws in solids. Firestone (1940) and Simons (1945) developed pulsed ultrasonic testing (UT) using a pulse-echo technique.

Inspection Methods 241

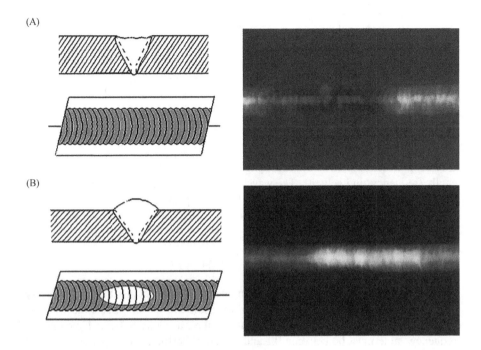

FIGURE 9.28 Inadequate welding reinforcement.

UT is a very short ultrasonic pulse-wave with center frequencies ranging from 0.1 to 15 MHz and occasionally up to 50 MHz penetrating into materials to detect internal flaws or to characterize materials. A common example is ultrasonic thickness measurement for measuring the webs and flanges thickness for the existing steel structure.

9.2.4.1 Wave Propagation

UT is based on time-varying vibrations in materials, which is generally referred to as acoustics. By applying ultrasonic sound to a specimen can determine its soundness, thickness, or other physical properties.

There are many types of waves of sound, including longitudinal waves, shear waves, surface waves, and in thin materials plate waves. The longitudinal and shear waves are the two modes of propagation most widely used in UT.

The longitudinal waves propagate in the longitudinal direction. For the transverse or shear wave, the particles oscillate at a right angle or transverse to the direction of propagation. Shear waves require an acoustically solid material for effective propagation, and, therefore, are not effectively propagated in materials such as liquids or gasses. Shear waves are relatively weak as compared with longitudinal waves. In fact, shear waves are usually generated in materials using some of the energy from longitudinal waves.

The sending and receiving waves of UT in the steel structure are different than in the concrete. As discussed before, the receiving and sending waves by one transducer are shown in Figure 9.29.

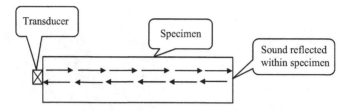

FIGURE 9.29 Sound propagation.

Among the properties of waves propagating in isotropic solid materials are wavelength, frequency, and velocity. The wavelength is directly proportional to the velocity of the wave and inversely proportional to the frequency of the wave.

The relation between the wave length, sound velocity, and frequency is as follows:

$$\text{Wave length } (\lambda) = \frac{\text{Velocity (V)}}{\text{frequency (f)}}$$

The wave length can be obtained by changing the frequency as the wave velocity is fixed value depending on the specimen materials. Based on ASNT, the smallest discontinuity you can detect by UT is about half the wave length (0.5λ). Note that the frequency value must be between 0.1 and 1 MHz (1 million cycles per second) and the wave velocity must be between 0.1 and 0.7 cm/us.

9.2.5 Reflection in Sound Wave

The wave velocity is constant through a given medium. There is a parameter called the acoustic impedance which is multiplying the wave velocity for a certain medium with the density of that medium.

Table 9.13 presents the velocity, density, and impedance in different medium.

Acoustic impedance is important in the determination of acoustic transmission and reflection at the boundary of two materials having different acoustic impedances. It is required for design of ultrasonic transducers and assessing absorption of sound in a medium.

Ultrasonic waves are reflected at boundaries where there is a difference in acoustic impedances of the materials on each side of the boundary. This difference in impedance is commonly referred to as the impedance mismatch. The greater the

TABLE 9.13
Acoustic Sound

Material	Velocity (cm/s)	Density (gm/cm³)	Impedance (gm/cm²-s)
Air	0.33×10^5	0.001	0.000033×10^6
Water	1.49×10^5	1.00	0.149×10^6
Aluminum	6.35×10^5	2.71	1.72×10^6
Steel	5.85×10^5	7.8	4.56×10^6

Inspection Methods

impedance mismatch, the greater the percentage of energy that will be reflected at the interface or boundary between one medium and another. In our cases we have two medium which is the steel and air.

The reflection and reason of attenuation are clearly shown in Figure 9.30. If reflection and transmission at interfaces is followed through the component, only a small percentage of the original energy makes it back to the transducer, even when loss by attenuation is ignored. For example, consider an immersion inspection of a steel block. The sound energy leaves the transducer, travels through the water, encounters the front surface of the steel, encounters the back surface of the steel, and reflects back through the front surface on its way back to the transducer. At the water steel interface (front surface), 12% of the energy is transmitted. At the back surface, 88% of the 12% that made it through the front surface is reflected. This is 10.6% of the intensity of the initial incident wave. As the wave exits the part back through the front surface, only 12% of 10.6 or 1.3% of the energy.

The reflection factor can be calculated from the following equation:

$$\text{Reflection factor (R)} = \left(\frac{Z_1 - Z_2}{Z_1 + Z_2}\right)$$

where Z is the acoustical impedance. If we apply the equation above for water to steel interface, the reflection value will be 88%.

When an ultrasonic wave passes through an interface between two materials at an oblique angle, and the materials have different indices of refraction, both reflected and refracted waves are produced. This also occurs with light, which is why objects seen across an interface appear to be shifted relative to where they really are. For example, if you look straight down at an object at the bottom of a glass of water, it looks closer than it really is. A good way to visualize how light and sound refract is

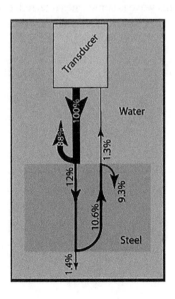

FIGURE 9.30 Reflection.

to shine a flashlight into a bowl of slightly cloudy water noting the refraction angle with respect to the incident angle.

Refraction of sound wave and snell law's is presented in Figure 9.31, where V_{L1} is the longitudinal wave velocity in material (1) and V_{L2} is the longitudinal wave velocity in material (2).

The velocity of sound in each material is determined by the material properties (elastic modulus and density) for that material. Therefore, when the wave encounters the interface between these two materials, the portion of the wave in the second material is moving faster than the portion of the wave in the first material. It can be seen that this causes the wave to bend.

Snell's Law describes the relationship between the angles and the velocities of the waves. Snell's law equates the ratio of material velocities V_1 and V_2 to the ratio of the **sine's** of incident (θ_1) and refracted (θ_2) angles, as shown in the following equation.

$$\frac{\sin\theta_1}{V_{L1}} = \frac{\sin\theta_2}{V_{L2}}$$

Note that the wave is reflected at the same angle as the incident wave because the two waves are traveling in the same material, and hence have the same velocities. This reflected wave is unimportant in our explanation of Snell's Law, but it should be remembered that some of the wave energy is reflected at the interface.

When a longitudinal wave moves from a slower to a faster material, there is an incident angle that makes the angle of refraction for the wave 90°. This is known as the first critical angle. The first critical angle can be found from Snell's law by putting in an angle of 90° for the angle of the refracted ray as shown in Figure 9.31. At the critical angle of incidence, much of the acoustic energy is in the form of an inhomogeneous compression wave, which travels along the interface and decays exponentially with depth from the interface. This wave is sometimes referred to as a "creep wave." Because of their inhomogeneous nature and the fact that they decay rapidly, creep waves are

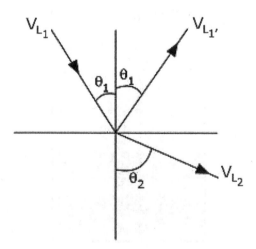

FIGURE 9.31 Reflection angles.

Inspection Methods 245

not used as extensively as Rayleigh surface waves in NDT. However, creep waves are sometimes more useful than Rayleigh waves because they suffer less from surface irregularities and coarse material microstructure due to their longer wavelengths.

9.2.6 Wave Interaction or Interference

The inspector should be familiar and have the full capability to understand everything about the discontinuities on the steel plate from the UT screen. Figure 9.32 presents the echo that will be presented in the UT screen.

There are three basic types of visual display: A-scan, B-scan, and C-scan. A-scan which is the time versus amplitude display which reveals a discontinuity on a cathode ray tube. The A-scan as shown in the following Figure 9.32 and the photo in Figure 9.33. For the B-scan is depending of using a cross sectional view of the testing

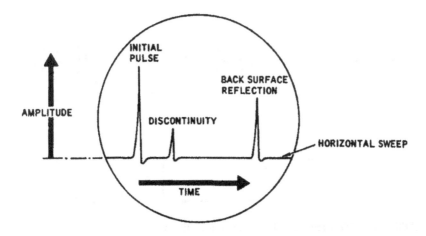

FIGURE 9.32 The shape of UT screen in case of discontinuity.

FIGURE 9.33 UT machine presenting first echo.

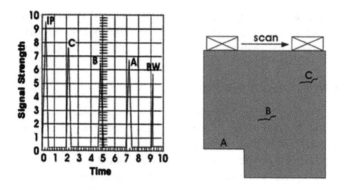

FIGURE 9.34 Reading of UT screen.

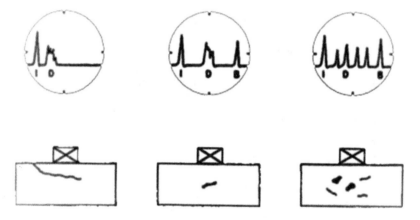

FIGURE 9.35 UT screen for different discontinuities pattern.

materials. For the C-scan is similar to the X-ray image which provide a plan view but the depth cannot be obtained. In most cases A-scan is used.

The A-scan presentation displays the amount of received ultrasonic energy as a function of time. The relative amount of received energy is plotted along the vertical axis and the elapsed time is presented in the x-axis, as shown in Figure 9.34 the location of crack B, C, and the reduction on thickness in A is presented by the following echo in the UT screen as shown in Figure 9.34. Note that the last echo BW is due to back wall echo.

For different patterns of cracks, the pulses on the UT screen will be different and should be well-known by the inspector. Examples of different discontinuities with different echo shape are presented in Figure 9.35.

9.2.7 Penetration Test

The penetrate test is a cheaper and easily-available technique to define the discontinuities on the surface of the steel and in the welding zone. This test depends on penetration of a liquid to revealing surface-breaking flaws by bleed-out of a colored or fluorescent dye from the flaw.

Inspection Methods

After applying the liquid by spray to the steel surface, it takes a period of time called the "dwell." Excess surface penetrant is removed and a developer applied. Contrast penetrants require good white light. Fluorescent penetrants need to be used in darkened conditions with an ultraviolet "black light," but it does not used to examine steel structure.

The surface preparation is one of the most critical steps of a liquid penetrant inspection. The surface must be free of oil, grease, water, or other contaminants that may prevent penetrant from entering flaws. The sample may also require etching if mechanical operations such as machining, sanding, or grit blasting have been performed. These and other mechanical operations can smear metal over the flaw opening and prevent the penetrant from entering. Once the surface has been thoroughly cleaned and dried, the penetrant liquid is applied by spraying in most cases but there are other types such as brushing or immersing the part in a penetrant bath in some industrial applications.

The penetrant is left on the surface for a sufficient time period to allow as much penetrant as possible to be drawn from or to seep into a defect. Penetrant dwell time is the total time that the penetrant is in contact with the part surface. Dwell times are usually recommended by the penetrant producers. The times vary depending on the application, penetrant materials used, the material, the form of the material being inspected, and the type of defect being inspected for. Minimum dwell times typically range from 5 to 60 min. Generally, there is no harm in using a longer penetrant dwell time as long as the penetrant is not allowed to dry. The ideal dwell time is often determined by experimentation and may be very specific to a particular application.

After that, a thin layer of developer is applied to the sample to draw penetrant trapped in flaws back to the surface where it will be visible. Developers come in a variety of forms that may be applied by dusting (dry powdered), dipping, or spraying (wet developers).

The developer is allowed to stand on the part surface for a period of time sufficient to permit the extraction of the trapped penetrant out of any surface flaws. This development time is usually a minimum of 10 min. Significantly longer times may be necessary for tight cracks.

Finally, the inspection is then performed under appropriate lighting to detect indications from any flaws which may be present. The final step is to clean the part surface to remove the developer from the parts that were found to be acceptable.

9.2.7.1 Advantages and Disadvantages of Penetrant Testing

To decide when to use this technique or not, it is important to know clearly both advantages and disadvantages of the liquid penetrant inspection. The primary advantages and disadvantages as described in dry penetrate method in ASTN (1980) when compared to other NDE methods are summarized below.

Primary Advantages
- The liquid penetrate test is highly sensitive to small surface discontinuities.
- The method has few material limitations, i.e., metallic and nonmetallic, magnetic and nonmagnetic, and conductive and nonconductive materials may be inspected.

- Large areas and large volumes of parts or materials can be inspected rapidly and at low cost it is used in the case of tanks inspection.
- Parts with complex geometric shapes can be routinely inspected.
- Indications are produced directly on the surface of the part and constitute a visual representation of the flaw.
- It is very portable in Aerosol spray as a penetrant material.
- Penetrant materials and associated equipment are relatively inexpensive.

Primary Disadvantages
- It examines only surface breaking defects.
- It can be used only on materials with a relatively nonporous surface can be inspected, so it is recommended in steel structure.
- The surface precleaning is critical since contaminants can mask defects.
- The inspector must have direct access to the surface being inspected.
- Surface finish and roughness can affect inspection sensitivity.
- Multiple process operations must be performed and controlled.
- It is essential of post cleaning of acceptable materials after inspection.
- Chemical handling and proper disposal is required.

9.2.7.2 Penetrant Testing Materials

The penetrates that are using these days are made to produce the level of sensitivity desired by the inspector. The following is the penetrate characteristics which are important to the inspector:

- it should spread easily over the surface of the material being inspected to provide complete and even coverage.
- it should be drawn into surface breaking defects by capillary action.
- it remain in the defect but can be removed easily from the surface of the part.
- it remains fluid so it can be drawn back to the surface of the part through the drying and developing steps.
- it can be highly visible or fluoresce brightly to produce easy to see indications.
- it will not be harmful to the material being tested or the inspector.

Figure 9.36 presents the shape of a discontinuity of applying the penetrate. All penetrant materials do not perform the same and are not designed to perform the same. Penetrant manufactures have developed different formulations to address a variety of inspection applications. Some applications call for the detection of the smallest defects possible and have smooth surfaces where the penetrant is easy to remove. In other applications, the rejectable defect size may be larger and a penetrant formulated to find larger flaws can be used. The penetrants that are used to detect the smallest defect will also produce the largest amount of irrelevant indications.

Penetrant materials are classified in the various industry and government specifications by their physical characteristics and their performance. Aerospace Material Specification (AMS) 2644, Inspection Material, Penetrant, is now the primary

Inspection Methods 249

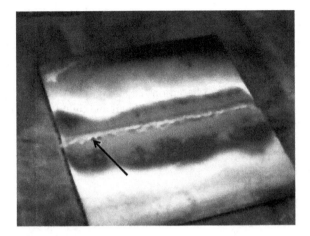

FIGURE 9.36 The shape of discontinuities in liquid penetrate.

specification used in the USA to control penetrant materials. Historically, Military Standard 25135, Inspection Materials, Penetrants, has been the primary document for specifying penetrants but this document is slowly being phased out and replaced by AMS 2644. Other specifications such as ASTM 1417, Standard Practice for Liquid Penetrant Examinations, may also contain information on the classification of penetrant materials but they are generally referred back to MIL-I-25135 or AMS 2644.

Penetrant materials come in two basic types. These types are listed below:

- Type 1: Fluorescent Penetrants
- Type 2: Visible Penetrants

Fluorescent penetrants contain a dye or several dyes that fluoresce when exposed to ultraviolet radiation. Visible penetrants contain a red dye that provides high contrast against the white developer background. Visible penetrants are also less vulnerable to contamination from things such as cleaning fluid that can significantly reduce the strength of a fluorescent indication.

On the other hand, penetrants are then classified by the method used to remove the excess penetrant from the part. The four methods are listed below:

- Water Washable
- Post-Emulsifiable, Lipophilic
- Solvent Removable
- Post-Emulsifiable, Hydrophilic

Water washable penetrants can be removed by rinsing with water alone. These penetrants contain an emulsifying agent (detergent) that makes it possible to wash the penetrant from the part surface with water alone. Water-washable penetrants are sometimes referred to as self-emulsifying systems. Post-emulsifiable penetrants come in two varieties: lipophilic and hydrophilic. In post-emulsifiers, lipophilic systems, the penetrant is oil soluble and interacts with the oil-based emulsifier to make removal possible. Post-emulsifiables, hydrophilic systems (Method D), uses an emulsifier that is a water-soluble detergent that lifts the excess penetrant from the surface of the part with a water wash. Solvent removable penetrants require the use of a solvent to remove the penetrant from the part.

Penetrants are then classified based on the strength or detectability of the indication that is produced for a number of very small and tight fatigue cracks. The five sensitivity levels are shown below:

- Level ½: Ultra Low Sensitivity
- Level 1: Low Sensitivity
- Level 2: Medium Sensitivity
- Level 3: High Sensitivity
- Level 4: Ultra-High Sensitivity

The major US government and industry specifications currently rely on the US Air Force Materials Laboratory at Wright-Patterson Air Force Base to classify penetrants into one of the five sensitivity levels. This procedure uses titanium and Inconel specimens with small surface cracks produced in low cycle fatigue bending to classify penetrant systems. The brightness of the indication produced is measured using a photometer. The sensitivity levels and the test procedure used can be found in Military Specification MIL-I-25135 and AMS 2644, Penetrant Inspection Materials.

An interesting note about the sensitivity levels is that only four levels were originally planned. However, when some penetrants were judged to have sensitivities significantly less than most others in the level 1 category, the ½ level was created. An excellent historical summary of the development of test specimens for evaluating the performance of penetrant materials can be found in the following reference.

9.2.7.2.1 Penetrants

The industry and military specifications that control penetrant materials and their use all stipulate certain physical properties of the penetrant materials that must be met. Some of these requirements address the safe use of the materials, such as toxicity, flash point, and corrosiveness, and other requirements address storage and contamination issues. Still others delineate properties that are thought to be primarily responsible for the performance or sensitivity of the penetrants. The properties of penetrant materials that are controlled by AMS 2644 and MIL-I-25135E include flash point, surface wetting capability, viscosity, color, brightness, ultraviolet stability, thermal stability, water tolerance, and removability.

9.2.7.2.2 Developers

The role of the developer is to pull the trapped penetrant material out of defects and spread it out on the surface of the part so it can be seen by an inspector. The fine developer particles both reflect and refract the incident ultraviolet light, allowing more of it to interact with the penetrant, causing more efficient fluorescence. The developer also allows more light to be emitted through the same mechanism. This is why indications are brighter than the penetrant itself under UV light. Another

Inspection Methods

FIGURE 9.37 Using spray for developer, penetrate, and cleaner.

function that some developers perform is to create a white background so there is a greater degree of contrast between the indication and the surrounding background.

The AMS 2644 and Mil-I-25135 classify developers into six standard forms. These forms are listed below:

1. Form a: Dry Powder
2. Form b: Water Soluble
3. Form c: Water Suspendable
4. Form d: Nonaqueous Type 1 Fluorescent (Solvent Based)
5. Form e: Nonaqueous Type 2 Visible Dye (Solvent Based)
6. Form f: Special Applications

The developer classifications are based on the method that the developer is applied. The developer can be applied as a dry powder, or dissolved or suspended in a liquid carrier. Each of the developer forms has advantages and disadvantages.

Nonaqueous developers is the most traditional technique to be used specific in steel structure projects. Figure 9.37 presents the spray of nonaqueous for the cleaner, penetrant, and the developer. Nonaqueous developers are commonly distributed in aerosol spray cans for portability. The solvent tends to pull penetrant from the indications by solvent action. Since the solvent is highly volatile, forced drying is not required. A nonaqueous developer should be applied to a thoroughly dried part to form a slightly translucent white coating.

When it is not practical to conduct a liquid penetrant examination within the temperature range of 40°F–125°F (5°C–52°C), the examination procedure at the proposed lower or higher temperature range requires qualification of the penetrant materials and procedure.

9.2.8 Magnetic Particle Inspection

Magnetic particle inspection (MPI) is a nondestructive testing method used for defect detection. MPI is fast and relatively easy to apply, and part surface preparation is not as critical as it is for some other NDT methods. These characteristics make MPI one of the most widely utilized nondestructive testing methods.

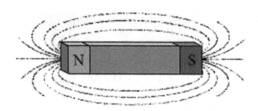

FIGURE 9.38 Magnet poles.

The MPI is depending in using magnetic fields and small magnetic particles (i.e., iron filings) to detect discontinuities on the surface of the materials. The only requirement from an inspectability standpoint is that the component being inspected must be made of a ferromagnetic material such as iron, nickel, cobalt, or some of their alloys. Ferromagnetic materials are materials that can be magnetized to a level that will allow the inspection to be effective.

In theory, MPI is a relatively simple concept. It can be considered as a combination of two nondestructive testing methods: magnetic flux leakage testing and visual testing. Consider the case of a bar magnet. It has a magnetic field in and around the magnet. Any place that a magnetic line of force exits or enters the magnet is called a pole. A pole where a magnetic line of force exits the magnet is called a north pole and a pole where a line of force enters the magnet is called a south pole. The poles of magnet are presented in Figure 9.38.

When a magnetic part has a crack as shown in Figure 9.38 it will formulate two different magnetic poles a north and south pole will form at each edge of the crack. The magnetic field exits the north pole and reenters at the south pole. The magnetic field spreads out when it encounters the small air gap created by the crack because the air cannot support as much magnetic field per unit volume as the magnet can. When the field spreads out, it appears to leak out of the material and, thus is called a flux leakage field.

If iron particles are sprinkled on a cracked magnet, the particles will be attracted to and cluster not only at the poles at the ends of the magnet, but also at the poles at the edges of the crack. This cluster of particles is much easier to see than the actual crack and this is the basis for MPI (Figure 9.39).

The first step in an MPI is to magnetize the component that is to be inspected. If any defects on or near the surface are present, the defects will create a leakage field. After the component has been magnetized, iron particles, either in a dry or wet suspended form, are applied to the surface of the magnetized part. The particles will be attracted and cluster at the flux leakage fields, thus forming a visible indication that the inspector can detect.

9.2.8.1 Magnetic Field Characteristics

Magnets come in a different shapes and the horseshoe (U) magnet is as shown in Figure 9.40 is one of the more common. The horseshoe magnet has north and south poles just like a bar magnet but the magnet is curved so the poles lie in the same plane. The magnetic lines of force flow from pole to pole just like in the bar magnet.

Inspection Methods 253

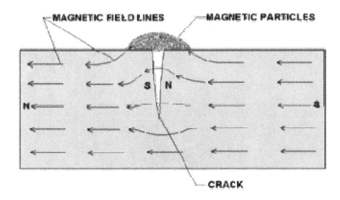

FIGURE 9.39 Magnet poles in metal cracks.

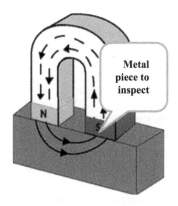

FIGURE 9.40 Direction of magnetic fields.

However, since the poles are located closer together and a more direct path exists for the lines of flux to travel between the poles, the magnetic field is concentrated between the poles.

If a bar magnet was placed across the end of a horseshoe magnet or if a magnet was formed in the shape of a ring, the lines of magnetic force would not even need to enter the air. The value of such a magnet where the magnetic field is completely contained with the material probably has limited use. However, it is important to understand that the magnetic field can flow in loop within a material. (See section on circular magnetism for more information) (Figure 9.41).

Magnetic lines of force have a number of important properties, which include:

- They seek the path of least resistance between opposite magnetic poles. In a single bar magnet as shown to the right, they attempt to form closed loops from pole to pole.
- They never cross one another.
- They all have the same strength.

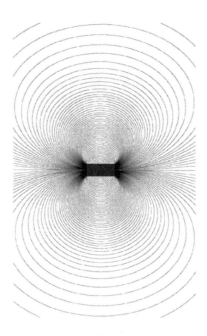

FIGURE 9.41 Magnetic field in horizontal plan.

- Their density decreases (they spread out) when they move from an area of higher permeability to an area of lower permeability.
- Their density decreases with increasing distance from the poles.
- They are considered to have direction as if flowing, though no actual movement occurs.
- They flow from the south pole to the north pole within a material and north pole to south pole in air.

9.2.9 Electromagnetic Fields

The magnetic field will formulate in circular form around the wire and that the intensity of the field was directly proportional to the amount of current carried by the wire. He also found that the strength of the field was strongest next to the wire and diminished with distance from the conductor until it could no longer be detected. In most conductors, the magnetic field exists only as long as the current is flowing. Note that in ferromagnetic materials the electric current will cause some or all of the magnetic domains to align and a residual magnetic field will remain.

To properly inspect a component for cracks or other defects, it is important to understand that the orientation between the magnetic lines of force and the flaw is very important. There are two general types of magnetic fields that can be established within a component.

A longitudinal magnetic field has magnetic lines of force that run parallel to the long axis of the part. Longitudinal magnetization of a component can be accomplished using the longitudinal field set up by a coil or solenoid. It can also be accomplished using permanent magnets or electromagnets.

Inspection Methods

A circular magnetic field has magnetic lines of force that run circumferentially around the perimeter of a part. A circular magnetic field is induced in an article by either passing current through the component or by passing current through a conductor surrounded by the component.

The type of magnetic field established is determined by the method used to magnetize the specimen. Being able to magnetize the part in two directions is important because the best detection of defects occurs when the lines of magnetic force are established at right angles to the longest dimension of the defect. This orientation creates the largest disruption of the magnetic field within the part and the greatest flux leakage at the surface of the part. As can be seen in the image below, if the magnetic field is parallel to the defect, the field will see little disruption and no flux leakage field will be produced.

An orientation of 45°–90° between the magnetic field and the defect is necessary to form an indication. Since defects may occur in various and unknown directions, each part is normally magnetized in two directions at right angles to each other. If the component below is considered, it is known that passing current through the part from end to end will establish a circular magnetic field that will be 90° to the direction of the current. Therefore, defects that have a significant dimension in the direction of the current (longitudinal defects) should be detectable. Alternately, transverse-type defects will not be detectable with circular magnetization.

There are a variety of methods that can be used to establish a magnetic field in a component for evaluation using MPI. It is common to classify the magnetizing methods as either direct or indirect.

9.2.9.1 Tools for Testing

In the case of steel structure section it is normal to use the prop magnetization as shown in Figure 9.42. By applying the hand rule theory, the direction of magnetization is as illustrated in the figure parallel to the welding line.

Table 9.14 presents the relation between the steel plate thicknesses and the ampere with the prod spacing (Table 9.14).

FIGURE 9.42 Magnet probes.

As we mentioned in the following section that half wave direct current provides a better powder mobility than DC. It also consumes less power and correspondingly produces a lower heat effect at the prod contact point. As per ASME V recommendation, to avoid arcing a remote control switch, which may be built into the prod handles, shall be provided to permit the current to be applied after the prods have been properly positioned.

The another regular method in the market which is the YOKE this machine will be used to magnetize a specimen longitudinally as shown in Figure 9.43. There is a coin around horseshoe magnet and it is made of soft, low retentivity iron which is magnetizing by a small coil wound around its horizontal bar. The machine also follows the right-hand rule to demonstrate the current flow. As per ASME V. this method shall only be applied to detect discontinuities that are open to the surface of the part.

To calibrate the yoke machine by applying the maximum lifting force of the AC yoke should be determined at the actual leg separation to be used in the examination as per ASME V. This may be accomplished by holding the yoke with a 10 lb (4.5 kg) ferromagnetic weight between the legs of the yoke and adding additional weights, calibrated on a postage or other scale, until the ferromagnetic weight is released. The lifting power of the yoke shall be the combined weight of the ferromagnetic material and the added weights, before the ferromagnetic weight was released. Other methods may be used such as a load cell.

TABLE 9.14
Recommended Current

	Thickness of Plate (inches)	
Prod Spacing (inches)	**Under ¾ inch (amperes)**	**¾ inch and Over (amperes)**
2–4	200–300	300–400
Over 4 to less than 6	300–400	400–600
6–8	400–600	600–800

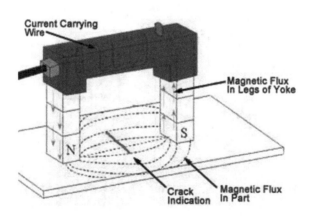

FIGURE 9.43 Yoke tool for MPI.

9.3 OFFSHORE STRUCTURE INSPECTION

The offshore structure inspection is the same methods of steel structure but in this case we will do the inspection under water, so it will need another technique. For offshore structure, we will use a visual inspection, UT inspection to measure the corrosion thickness and we can use MPI under water also with special materials. For the tubular member, we use a flooded member detection (FMD) technique. Using this method we can define if there is water inside the bracing pipe, meaning that there is a crack for the two ends or one end for the tubular joint. In addition to that, the cathodic protection measure will be done also. These techniques can be done by using divers or can be done by using remote-operating vehicle (ROV) (Figure 9.44).

It can use magnetic particle tests in underwater survey as shown in Figure 9.45. It is required to define if there are any cracks in the tubular joints.

FIGURE 9.44 ROV subsea inspection.

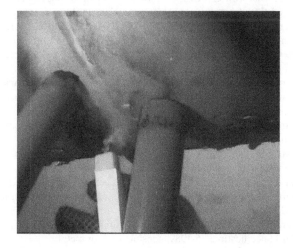

FIGURE 9.45 Underwater MPI.

BIBLIOGRAPHY

ACI 228.1R89, In-place methods or determination of strength o concrete, ACI manual of concrete practice, part 2: construction practices and inspection pavements, 25 pp., Detroit, MI, 1994.

American Society of Nondestructive Testing, ASNT, Radiographic method manual, 1980.

American Society of Nondestructive Testing, ASNT, Dry penetrate training manual, 1980.

American Society of Nondestructive Testing, ASNT, Magnetic particle test training manual, 1980.

American Society of Nondestructive Testing, ASNT, Ultrasonic method training manual, 1980.

ASTM C188-84, Density of hydraulic cement.

ASTM C114-88, Chemical analysis of hydraulic cement.

ASTM C183-88, Sampling and amount of testing of hydraulic cement.

ASTM C349-82, Compressive strength of hydraulic cement Mortars.

ASTM C670-84, Testing of building materials.

ASTM C142-78, Test method for clay lumps and friable particles in aggregate.

ASTM D1888-78, Standard test method for particulate and dissolved matter, solids or residue in water.

ASTM D512-85, Standard test method for chloride ion in water.

ASTM D516-82, Standard test method for sulfate ion in water.

BS 410-1:1986, Spec. for test sieves of metal wire cloth, 2000.

Bs EN933-1, Tests for geoetrical properties of are gates, determination of particle size distribution. Sieving method, 1997.

BS 812 Part 103-1985, Sampling and testing of mineral aggregate sands and fillers.

BS 882: 1992.

Ajdukiewicz, A.B., and Kliszczewicz, A.T., 1999, Utilization of recycled aggregatesin HS/HPC, *5th International Symposium on Utilization of High Strength/High Performance Concrete*, Sandefjord, Norway, June 1999, V. 2, pp. 973–980.

Ajdukiewicz, A.B., and Kliszczewicz, A.T., 2000, Properties and usability of HPC with recycled aggregates, *Proceeding, PCI/FHWA/FIP International Symposium on High Performance Concrete*, Sep. 2000, Orlando, Precas/Prestressed Concrete Institute, Chicago, 2000, pp. 89–98.

Ajdukiewicz, A.B., Kliszczewicz, A.T., 2002, Behavior of RC beams from recycled aggregate concrete, *ACI Fifth International Conference*, Cancun, Mexico.

Di Niro, G., Dolara, E., Cairns R., 1998, The use of recycled aggregate concrete for structural purposes in prefabrication, *Proceedings, 13th FIP Congress "Challenges for Concrete in the Next Millennium"*, Amsterdam, June 1998; Balkema, Rotterdam-Brookfield, 1998, V. 2, pp. 547–50.

Egyptian Standard Specification: 1947-1991, Method of taking cement sample.

Egyptian Standard Specification: 2421-1993, Natural and mechanical properties for cement. Part 2: define cement finening by seive No.170.

Egyptian Standard Specification: 2421-1993, Natural and mechanical properties for cement. Part 2: define cement finening by using blain apparatus.

Egyptian Standard Specification: 2421-1993, Natural and mechanical properties for cement. Part 1: define cement setting time.

Egyptian Standard Specification: 1450-1979, Portland cement with fines 4100.

Egyptian Standard Specification: 1109-1971, Concrete aggregate from natural resources.

ISO 6274-1982, Sieve analysis of aggregate.

Egyptian Standard Specification: 262-1999, Steel reinforcement bars.

Egyptian Standard Specification: 76-1989, Tension tests for metal.

ECP203, Egyptian code for design and execute concrete structures: part 3 laboratory test for concrete materials, 2003.

Kasai, Y., (ed.), 1988, Demolition and reuse of concrete and masonry reuse of demolition waste, *2nd International Symposium RILEM*, Building Research Institute and Nihon University, Tokyo, Nov. 1988, Chapman and Hall, London and New York, p. 774.

Mukai, T., and Kikuchi, M., 1988, Properties of reinforced concrete beams containing recycled aggregate, *Proceedings, 2nd International Symposium RILEM "Demolition and Reuse of Concrete and Masonry"*, Tokyo, Nov., Chapman and Hall, London and New York, V. 2, 1988, pp. 670–79.

Murphy, W.E., discussion on paper by Malhotra, V.M., 1977. Contract strength requirements-core versus in situ evaluation, *Journal of the American Concrete Institute*, Vol. 74, No. 10, pp. 523–25.

Neville, A.M., Properties of concrete, PITMAN, 1983.

Plowman, J.M., Smith, W.F., and Sheriff, T., 1974, Cores, cubes, and the specified strength of concrete, *The Structural Engineer*, Vol. 52, No. 11, pp. 421–26.

Salem, R.M., and Burdette, E.G., 1998, Role of chemical and mineral admixtures on physical properties and frost – resistance of recycled aggregate concrete, *ACI Materials Journal*, Vol. 95, No. 5, pp. 558–63.

Van Acker, A., 1997. Recycling of concrete at a precast concrete plant. FIP Notes, Part1: No. 2, 1997, pp. 3–6; Part 2: No. 4, 1997, pp. 4–6.

Yuan, R.L. et al., 1991, Evaluation of core strength in high-strength concrete, *Concrete International*, Vol. 13, No. 5, pp. 30–34.

10 Repair of Concrete, Steel, and Offshore Structures

10.1 INTRODUCTION

As we explained in the previous chapter, the collapse of a building refers to a complete failure of the building as it reaches the structure to the ground. The failure of a building means that it fails to be used or do its function correctly. We should repair or strengthen the structure to withstand the required load in the case of a structure failure or a partial structure failure. In some cases, one can also do a load reduction to reduce the structure probability of failure and increase its reliability.

In most cases, the repair and retrofit of a concrete structure is very complicated than executing a new concrete structure, as new constructions do not need competent engineers with vast experience and can be handled by junior engineers with reasonable amounts of experience, for example, 2 or 3 years. On the other hand, the repair of concrete structure is more challenging and needs competent engineers and consultants as well.

The building already exists and you need to define the new solutions to the problem that must be matched with the nature of the building and owner requirement while satisfying the safety and economic requirements.

The repair process for reinforced concrete structures is very important and dangerous. Extensive care must be taken in choosing the suitable repair methods and tools. It is worth to mention that many building collapses during the repairing process.

I usually give the example of a young man and an old man. If both of them go to the doctor, the young man will have no problem to the medicines but the old man may have a problem to the medicine, and it may cause side-effects in case of wrong medicine. The young man will recover but the old man may die "Ce La vie".

The structures that have not considered in the design phase or execution phase the code and specifications precautions to protect their structure or not taken into account in the design phase the necessary protection method or in contrary may it had been chosen the protection method but the way of execution or follow is bad.

In all the previous cases, the results will be the same, which is the structure deterioration and this will become obvious by cracks and falling of concrete cover of the different reinforced concrete elements of the structure.

When the cracks are present and the falling of concrete cover occurs, the situation becomes very critical. However, reduction occurs in a cross-sectional area of the concrete member, which requires speed up the work of repair as soon as possible to avoid the structural condition to be worst and more likely to collapse with time. As we discussed in the previous chapter, the probability of failure is very high.

The first and often most important step of repairing damaged or deteriorated concrete is to correctly determine the cause of damage. This depends entirely on the assessment of structure and the answer to the following questions:

- What is the reason of corrosion?
- Are the cracks and the deterioration of structure increasing?
- What is the expectation of the extent of deterioration of concrete and extension along the corrosion in the steel?
- What is the impact of the current and expected future deterioration to the safety of the structure?

Answering these questions will need to use the assessment methods of structure and different measurement methods to decide the cause of corrosion and the deterioration of structure in present and in future (see Chapters 8 and 9 for more details).

It is worth noting that the process of repair and restoration of concrete structures, from the assessment phase to the execution phase of structure, need high levels of experience. However, the simplest stage, such as the evaluation stage, when it is performed by inexperienced engineers, may lead to the wrong choice of repair method, and this is a critical issue.

When assessing the structures, there are several strategies for repair mentioned in a special report by RILEM (1994) that clarify the various repair strategies used in most cases.

1. Re-establishment of the deteriorated member.
2. Comprehensive repair of the concrete member to regain its ability to withstand the full loads.
3. Repair of a particular portion of the concrete member and follow-up by continuous periods of time.
4. Strengthening of the structure by an alternative system to bear part of the loads.

In most cases, the strategy of repair is either a comprehensive repair of the concrete member or a partial repair of the concrete member. These strategies are common in the rehabilitation of concrete and depend on the structural system, external environmental factors, and the degree of structural degradation.

10.2 CONCRETE REPAIR PROCEDURE

The first important step is to choose an experienced contractor who has a track record for doing such work of repair as it needs a specific experience from the engineers, technician, and also the labor.

There are several regular steps in the repair of all structures exposed to corrosion. The first, and very critical, step is to strengthen the structure by performing a structural analysis and deciding on the suitable location of the temporary support.

The second step is to remove the cracked and delaminated concrete. It is important to clean the concrete surface and also cleaning the steel bars by removing rust from it.

Concrete, Steel, and Offshore Structures 263

After removing rust by brush or sandblasting, the steel bars should be painted with epoxy coating or changed the steel bars by new one. Then new concrete can be poured. The final step is to paint the concrete member with concrete surface coating as external protection.

This is briefly the repair process. These steps will be explained in detail in the following sections.

Concrete strengthening is considered one of the dangerous and important steps, and is necessary for the repair. The temporary support selection depends on the following:

1. Evaluation of the state of whole structure.
2. How to transfer loads in the building and its distribution.
3. The volume of repair that will be done.
4. The type of the concrete member that will be repaired.

Therefore, as mentioned earlier, the repair process must be carried out by a structural engineer with a high degree of experience with the repair process, the capability to perform the structural analysis, and a powerful knowledge of the load distribution in the structure, according to the kind of repair that will be completed, because it often follows the repair process of breaking the defective concrete.

Therefore, any engineer has the responsibility to choose the right way to optimize the process of crushing and of determining the ability of members to carry the loads that will be transferred to them.

Therefore, the responsible engineer should design the temporary supports based on the obtained data and the previous analysis and take caution in the phase of execution of temporary supports.

It is worth mentioning that choosing the ways of removing the defected parts will be based on the nature of the concrete member in the building as a whole. Any member of breaking concrete has a detrimental impact on neighboring members, because the process of breaking will produce a high level of vibration.

The temporary members must be strong and should be designed in an appropriate manner to withstand loads transported from the defective members with easily and safely. You can imagine that the entire pool of structure depends on the design and execution of the temporary supports and their ability to bear the loads with adequate safety.

10.2.1 Removing Concrete Cover

Really, it is very critical steps as the structure strength shall be reduced very gradually, so you shall check and guarantee that the temporary support that the contractor is using is done be calculation and considering all the transferee loads.

As an example, there was a repair under progress in multistory building for the columns and as the contractor is not professional, he removed all the columns concrete covers in the same time to utilize the same labor and reduce cost as the result of this act is the complete collapse of the building.

There are several ways to remove the part of the concrete member that has a cracked surface and steel corrosion on the surface. These methods of removing the delaminated concrete depend on the ability of the contractor, specifications, the cost of the breakers, and the state of the whole structure.

The selection of the concrete cover demolishing methods based on the causes of corrosion if it is due to carbonation or chlorides and in addition to that it must consider if it is proposed to perform cathodic protection in future. In this situation, the demolishing work will be on the falling concrete cover and clean and remove all the delaminated concrete and, also the concrete parts that in its way to be cracked and then pouring high strength and non-shrinkage mortar.

If the corrosion in the steel reinforcement is a result of chloride propagation into the concrete, most specifications recommend removing about 25 mm behind the steel and therefore to make sure that the most concrete on the steel had no traces of chlorides after the repair process. The difference between good repair and bad repair procedure is shown in Figure 10.1.

The difference in the procedure of removing the delaminated concrete is due to the difference in the causes of corrosion. Therefore, a careful study to assess the state of structure and the causes of corrosion is very important to obtain high quality after the repair process. The evaluation process is the same as diagnosis of an illness. If there are many mistakes in the diagnosis, the repair process will be useless in addition to waste of time and money.

It is required to define the work procedure and quantity of concrete that will be removed. This step is considered one of the fundamental factors for designing and installing wooden pillars of the building to be used during execution. Therefore, the work plan must be clear and accurate for all the engineers, foramen, and workers who participate in the repair process.

After completing the building assessment and designing of supports and ties, as well as ensuring the presence of a competent staff, there may be some information that is not available, for example, the construction procedure for this building, workshop drawings, or specification that follows when the building is constructed. Therefore, the chances of risk are still high. The only factor that can help us in reducing this risk in spite of the steel corrosion is the increase in concrete strength with time. However, this compensation is within a limit as the steel is carrying most of the stress and the risk will be very high in the case of spalling the concrete cover due to reduction in concrete cross-sectional dimensions.

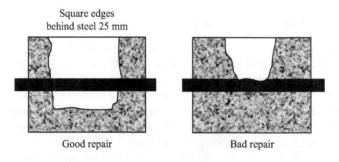

FIGURE 10.1 Differences between good and bad repair.

Concrete, Steel, and Offshore Structures

It is necessary and important to remove concrete for a distance greater than the volume required for removal of defective concrete, so that we can properly reach the steel, because it will be important later in the repair process.

There are several methods commonly used for breaking and removing the defective concrete. These will be explained in the following sections.

10.2.1.1 Manual Method

One of the simplest, easiest, and cheapest ways is to use a hammer and chisel. By using drawings, specialists prefer to remove defective concrete. It is too slow as compared with the mechanical methods.

However, we should note that the mechanical methods produce high noise and vibration and need special requirements and trained labor.

Using this manual method, it is difficult to spare concrete behind the steel. The method used in the case of small spaces and is used preferably in the event of corrosion due to carbonation and attacking chlorides from outside, which do not required breaking concrete behind the steel.

It is worth to mention that any worker can manually break the concrete, but you must choose workers who have worked before in repair, as they must be sensitive in breaking the concrete to avoid causing cracks to the adjacent concrete members.

10.2.1.2 Pneumatic Hammer Methods

These hammers work by compressed air and their weight is between 10 and 45 kg. If they are used on the roofs or walls, their weight will be about 20 kg. They need small power unit to do the job attached with it, but in large areas it may require a separate, bigger air compressor.

This machine needs properly trained workers. It is worth mentioning that the compressed air hammers have a fewd initial costs, of which a few were discussed by the Strategic Highway Research Program (SHRP), and based on Vorster et al. (1992), a research program of highways that the terms of the contract are governed by the contractor, which will use breaking machine.

In the case a small area to be removed, the use of pneumatic hammers is economical than in the case of large areas, as for large areas it is preferred to use water gun, which will be illustrated in the following section.

But in case of the client who does not specify the area that need to be removed in the tender the cost calculation will be in square meter. In this case, the risk will be low because the machine's initial cost is little, so in this situation it is preferred to use pneumatic hammer.

The performance rates of the breakers are about 0.025–0.25 m^3/h using hammers with and weighing 10–45 kg, respectively.

Figure 10.2 shows the summary of the pneumatic hammers used in small areas of concrete to be removed. It does not prefer to be used in removing a large area of concrete.

10.2.1.3 Water Jet

This method has been commonly used in the 1970s since its presence in the market. It relies on the existence of water at the worksite, depending on the removal of a

FIGURE 10.2 Using pneumatic hammer to remove wall concrete cover.

suitable depth of concrete in a large area. It removes fragmented concrete, cleans steel bars, and also removes part of the concrete behind the steel bars, as shown in Figure 10.3.

The water jet is used manually through an experienced worker who has dealt with a hose, which is pushing water under high pressure or perhaps through a mechanical arm as the manual needs to apply very high safety precautions to the worker who use it and the site around it.

The used water must not have any materials that affect the concrete as high chloride ions; for example, in general it must be potable water.

Water gun is consisting of a diesel engines, pressure pump and connected by hose bear high pressure resistance of water and water pressure up at the gun nozzle to 300–700 kg/cm^2. At least, 400 kg/cm^2 is required to cut the concrete and the rate of water consumption is about 50 L/min.

The performance rate of water jet to break is 0.25 m^3/h—in the case of the use of a small pump—and can reach up to about 1 m^3/h in the event of the use or two pumps or one big pump.

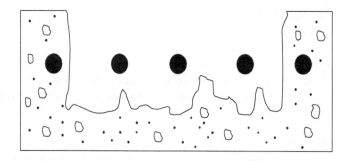

FIGURE 10.3 Shape of delaminated concrete by water jet.

10.2.1.4 Grinding Machine

It is used to remove concrete cover in the case of large flat surfaces as an example of that is the bridges deck, as shown in Figure 10.4, but it must be done cautiously so that the process of breaking do not reach the steel. As in the case any contact between the grinding machine and steel reinforcement it will cut the steel bars and damage the machine.

This method of breaking the delaminated concrete cover is little used in the United Kingdom because they often put insulation film to prevent the water, but used sparingly in the United States bridges.

The grinding machine is usually used after the water gun or the pneumatic hammer to obtain final concrete breakdown around and under the steel reinforcement.

We must therefore take into account whether the thickness of the concrete cover is equal. The rate of removal of the concrete by this machine is very fast; it removes about 1 m^3/min and its cutting part is 2 m in width.

10.2.2 CLEAN CONCRETE SURFACE AND STEEL REINFORCEMENT

This phase removes any remaining broken concrete with a process of cleaning and at the same time, the process of assessing the steel and cleaning up and removing corrosion from the roof takes place.

The stage of preparing a surface by pouring the new concrete is one of the most important stages of the repair process. as the process of repair, if they were in a good regardless of the cost, so we have prepared the main factor in the process of the bond between the new concrete and old concrete.

Before applying the primer coating, which provides the bond between the existing old concrete and the new concrete for repair, the concrete surface must be well prepared and this is according to the used materials.

In all cases, the concrete surface must be clean and does not contain any oils or broken concrete or soil or lubricants. If any of these elements is present, the surface must be cleaned completely through the sandblasting, water, or cleaning using brushes done manually.

FIGURE 10.4 Grinding machine.

This stage is very important and very necessary, regardless of the type of material used to bond the new concrete with the old. Neglecting the preparation of the surface might affect the repair process as a whole because this stage of repair is the less expensive stage in the repair process.

We do not need to prepare a surface when using a water gun to remove the delaminated concrete, as the surface will be wet enough and will be clean after the crushing process. It is beneficial to use a water gun because it uses air pressure to clean the surface of concrete and remove any growth and concrete fragmented used as supplement for the surface material used for linking with new concrete.

10.2.3 Concrete Repair Process

When using cement mortar or concrete in the repair, the concrete surface will be sprayed by water until it is saturated. The stage of saturation can be reached by spraying water on the surface for 24 h or through wetting burlap. Water spraying must be stopped and the burlap should be removed for about an hour to 2 h, depending on the weather conditions. Then surface is dried and coated with a mixture of water and cement, which is called "slurry." The material will be applied by brush and also can use an epoxy coating as an adhesive between new and old concrete and must follow the specifications and precautions which are stated by the manufacturer.

The preparation of the concrete surface to achieve better bond between the old and new concrete must be in line with the American specifications ACI 503.2-79, which states that surfaces that will receive epoxy compound applications must be given careful attention as the bonding capability of a properly selected epoxy for a given application is primarily dependent on proper surface preparation. Concrete surfaces to which epoxies are to be applied must be newly exposed and free from loose and unsound materials. All surfaces must be meticulously cleaned, as dry as possible, and at proper surface temperature at the time of epoxy application.

The epoxy materials used for bonding should be in line with the ASTM C881-78 code, which must be well-defined by the supplier specifications and identical to the different circumstances surrounding the project, particularly the temperature change that might occur in a resin as well as the nature of loads carried by the required repair member.

It is worth noting that during the execution of the repair process, the use of epoxies will be significant. Therefore, we must take into account the safety factor for workers who use such material. Therefore, the workers must wear their personal protective equipment (PPE) as gloves and special glasses for safety issues, as these materials are very harmful to skin and can cause many dangerous diseases if any one is exposed to it for a long time.

There are also some epoxies that are flammable at high temperatures, and that must be taken into account during storage and operation.

After removing the concrete covers and cleaning the surface, the next step is to evaluate the steel bar diameter.

If it was found that the cross-sectional areas of the steel bars have a reduction equal or more than 20%, it requires to add additional reinforcing steel bars and before pouring new concrete, one must sure that the development length between the new bars and the old steel bars must be enough.

It is usually preferred to link the steel by drilling new holes in the concrete and connecting the additional steel on concrete by putting the steel bars in the drilled hole filled with epoxy. However, in most cases the steel bars are completely corroded and need to be replaced.

10.2.4 Beam and Slab Repair Process

In case of adding steel reinforcement bars to beams and slabs, prefer to connect the new steel bars with concrete by drilling new holes in the concrete and make the bond of the steel bars in the holes by using adhesive epoxies.

In case of repair beams, the additional steel bars shall be fixed into the column which support this beam. In case of slabs the steel bars is fixed in the sides of the beam that is supporting the slab as shown in Figure 10.4. The dowels will be fix to the beam side by making drilling holes with depth around 70–80 mm and the dowel will be fix in the holes by epoxy.

The depth of the hole may differ, depending on the type of the epoxy and the supplier's assumption and recommendations. The hole's diameter will be 40 mm higher than the bars diameter to ensure that the steel bars are fixed completely.

It is best to use 12 or 16 mm and it is preferred to increase the bar diameters and decrease their number to reduce the number of holes in the beams. The dowel is put in the drilled hole, as shown in Figure 10.5, with a length 50 times the bar's diameter, which overlaps the new steel bars and is fixed at both directions by putting small dowel from the slab in the intersection between the steel bars; these small dowels are also fixed to the slab by epoxy.

Most coastal cities have special creations by the architectural, and design is based on placing all the balconies facing the coast to see the sea and to increase the value of apartment units. At the same time, the balconies required distinctive structural elements in their design and execution, as it is a cantilever from structural point of view and the cantilever has the lower redundancy in the all structural elements, so it needs special attention in the case of repair and restoration.

In the case of repairing the balconies as it is traditional to have corrosion on the upper steel reinforcement, as it is the main steel particularly in the cantilever. The tiles will be removed, the concrete cover removed, and then the reduction on the upper steel reinforcement will be inspected, as we mentioned earlier.

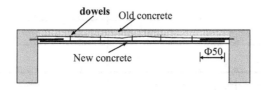

FIGURE 10.5 Steel reinforcement installation for slab repair.

If the condition of steel bars is bad, it will be necessary to add new upper steel reinforcement. Now put the upper steel bars to extend on the adjacent span by a length equal to 1.5 times the cantilever length, as shown in Figure 10.6.

The dowels will be installed vertically and fixed by the concrete slab with epoxy to a depth of 50 mm. The distance between the dowels is about 350 mm and these dowels are attached to the new main steel bars.

It is worth mentioning that in all repair processes, especially in cantilever as it is a point of weakness in terms of structure safety, one must pay attention to the design and execution to the wood form and temporary supports and the part of the building that will be overloaded during repair process by taking into consideration the structure system and stress distribution in different elements, the weak points in the building, the building mature, and the method of repair that will be applied.

As for the beam repair process, the work is the same as the slab repair process. However, the difference is in fixing the dowels in the columns as shown in Figure 10.7, and fixing the stirrups through making holes with a depth of approximately 50–70 mm (according to the epoxy manufacturer's recommendation). Epoxy is applied on both sides of the beam using to stabilize stirrups, about 8 mm per 200 mm, as shown in the above figure.

After that, as a normal procedure, epoxy coating is applied as an adhesion between old and new concrete and then the new concrete is poured manually, or by shotcreting.

As shown in Figure 10.8, the beam is strengthened by increasing the cross-sectional dimension and adding steel reinforcement.

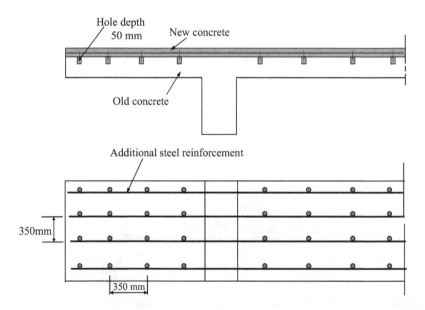

FIGURE 10.6 Cantilever repair.

Concrete, Steel, and Offshore Structures 271

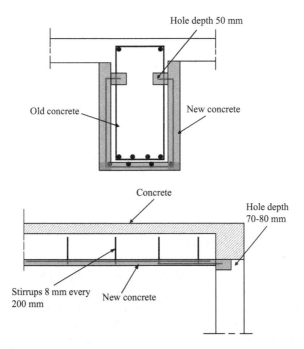

FIGURE 10.7 Beam repair.

FIGURE 10.8 Steel installed for beam repair.

10.2.5 Column Repair Process

As for the columns, the repair is performed as shown in Figure 10.9, which shows how casting on site, but we must know that the minimum distance allowed to cast concrete easily is about 100–120 mm from each side.

The dowel is fixed for the first floor as shown in Figure 10.9, which does not require drilling holes in the base, but making the legs of the steel bars to rest on the

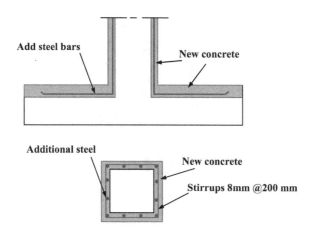

FIGURE 10.9 Repair of concrete columns.

foundation. The steel bars will be distributed around the column circumference as in the figure to cope with the reduction of steel reinforcement cross-sectional area due to corrosion and the steel cross-sectional area percentage to the concrete column cross-sectional area as the same percentage in the old column member and not less than as stated in the code. The exterior column addition of steel for repair is shown in Figure 10.10. The additional steel bars to repair the concrete pedestal foundation that carrying the pipeline are shown in Figure 10.11.

When water jet is used in the breakers, the steel will be cleaned and corrosion will be removed from it; however if it is not used in the water gun, the corrosion will be removed by sandblasting.

FIGURE 10.10 Adding steel for an exterior column.

Concrete, Steel, and Offshore Structures 273

FIGURE 10.11 Repair of pedestal carrying pipelines.

The old steel will be painted by using epoxies after cleaning steel bars completely, especially from the effects of chlorides. Slurry which is mixing between cement and water can be used, by painting the steel by this slurry to obtain the benefit for alkaline protection from the cement mortar to the steel reinforcement.

The improved cement slurry can be dried quickly so that it will be not effective in the repairs that require the installation of form after painting bars, but it works good in cases where the time between painting steel and pouring cement mortar is short. This time period should not exceed 15 min as stated by U.S. Army manual in 1995.

The repair of slabs and beams by adding steel is shown in Figure 10.12.

FIGURE 10.12 Slab and beam repair.

10.2.6 NEW CONCRETE PATCHES

There are some mixtures available on the market, which had been mixed in a certain way so that they can be used easily in the repairs for small areas; however, these mixtures are quite expensive.

In the case of a large surface area to be repaired, the mixture preparation and mixing would be onsite to be less costly, but this must be carried out by an expert.

In addition to these properties of concrete characteristics, using a pump for pouring or shotcrete needs a special design mix.

The ready-mix concrete factories can provide guarantee in case of corrosion due to carbonation but if it is due to chlorides they do not provide guarantee afraid from the presence of chloride after the repair process.

It is also worth to note that most manufacturers of materials used in mixing are field execution contractors, but when they supply materials only, they will provide all the information and technical recommendations for the execution, performance rates to calculate the required amount of the materials, and appropriate method of operation.

Worldwide, all companies operating in the construction field and dealing with chemicals must have a competent technical staff that can assist on-site or required to be under their supervision to avoid any error and to define the responsibility in case of any defects.

There are two types of materials used as a new mortar for repair: polymer mortar and cement mortar enhanced by polymer. Both will be described in the following sections.

10.2.6.1 Polymer Mortar

This mortar is ready to give the specific components of repair and give what is required from the ability to control the ease of operation and the influence as well as which can be controlled of the setting time.

This mortar provides a good bond with the existing old concrete, so it does not need any other to develop this cohesion. This mortar has high compression strength around 50–100 MPa and high tensile strength also.

Despite the many advantages of such mortar, its properties are different from those of the existing concrete as the polymers have a coefficient of thermal expansion equal to $65 \times 10^{-6}/°C$, corresponding to a concrete thermal expansion coefficient equal to $12 \times 10^{-6}/°C$. Moreover, the modulus of elasticity for mortar is much less than in the case of concrete. These differences in the properties lead to the presence of cracks as a result of generating internal stresses; however, they can be overcome by aggregate sieve grading to make the proportion of polymers less, which reduces the difference in the natural properties.

There are many tests to define this mortar and noncompliance with the concrete (e.g., ASTM C884-92 Standard Test Method for PCBs Comparability between Concrete and Epoxy-Resin over Lay).

10.2.6.2 Cement Mortar

Polymer mortar gives a physical protection to the steel reinforcement; however, the cement mortar gives passive protection as they increase the alkalinity around the steel reinforcement. The trend now is to use cement mortar for repair of damages that result of corrosion, as it has the same properties as the existing concrete and it gives passive protection from corrosion.

The usage of some polymers additives to the cement mortar through liquid or powder improves some of the properties of cement mortar (Ohama, 1995; ACI Committee 548, 1994), increases the flexural resistance, increases the elongation, reduces the water permeability, increases the bond between the old and new concrete, and increases the effectiveness of its operation.

The polymer used is identical to the cement mortar mixture and the properties that control the polymers in the cement mortar are stated in ASTM C1059-91 specification for latex agents for bonding fresh to hardened concrete.

Silica-fume can be used with mixture, as well as super plasticizers to improve the properties of the used mortar and reduce shrinkage.

10.2.7 Execution Methods

There are several ways to implement the repair process. These are entirely dependent on the type of structure member to be repaired and the materials used in the repair.

These methods are as follows:

10.2.7.1 Manual Method

They are used in most cases, particularly in the repair of small spaces and it is often used flawed in addressing relevant.

Casting way at the site. This is done by making a wooden form and then pouring concrete in the damaged part (Figure 10.13) of concrete columns or vertical walls.

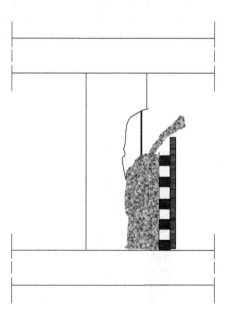

FIGURE 10.13 Casting concrete onsite.

One must consider an appropriate distance to cast the concrete in the wooden form. This efficient method is commonly used and does not require expensive equipment. However, users need to be experienced in the fabrication and installation of the wooden form and the casting procedures.

10.2.7.2 Grouted Pre-placed Aggregate

As shown in Figure 10.14, the aggregate should be placed with gap grading in the area to be repaired. The next step is to make a grouting fluid by injecting it inside the aggregate by a pipe with pump to fill the gap between the aggregates by the grout. This method of repair is used in repairing the bridge supports and other special application as in the marine structure.

This method is used in terms of special equipment as the pipe injection, pump, and other special miscellaneous equipment. Therefore, one can conclude that this method is used by private companies with high potential.

10.2.7.3 Shotcrete

They are used in the case of large surfaces to be repaired, but considers that the mix and components of concrete suitable for the use in shotcrete as they need a special additives and specifications.

As shown in Figure 10.15, the health and safety precaution must be carefully followed in this method for the worker who is using the shotcrete as they contain some polymer resins, as it contains polymers and special additives.

In the concrete mix design, the nominal maximum coarse aggregate size must be defined to be suitable to the shotcrete equipment's nozzle and pump to avoid any problem during casting the concrete.

Complete member casting. They are used when total reconstruction is required for the concrete member whose steel reinforcement bars have been depleted due to corrosion and therefore it is necessary to pour concrete for the complete member with full depth.

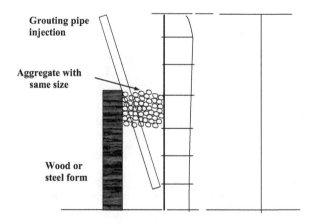

FIGURE 10.14 Injected pre-placed aggregates.

Concrete, Steel, and Offshore Structures 277

FIGURE 10.15 Shotcrete in concrete repairing.

It is frequently used in the repair of bathrooms as in some cases the concrete slab will be in a very severe bad condition, so the ordinary repair procedure will not be efficient. In this case, all the concrete slabs will be demolished, a new steel reinforcement installed, and then pour the concrete slab as shown in the figure. Before deciding to take the whole building condition must be taken into account. This method is usually easy to apply in the bathroom slab, as in residential building, the bathroom slab is usually designed as simply supported as it drops around 100 mm for plumping pipes.

As shown in Figure 10.16, this slab is deteriorated so it will be demolished and a new slab will be poured.

In any steel storage tank for oil or water, there will be a ring beam underneath the shell of the tank. The following figures present a case study for repairing reinforced concrete ring beam supporting steel tanks containing oil.

FIGURE 10.16 Pouring a new slab.

The ring beam is structurally carrying a tensile strength due to an earth load affecting the beam, which puts the ring beam under tension. Corrosion of the steel reinforcement bars means reduction on the capacity of the ring beam. So, the best solution is to increase the steel bars to overcome the reduction on the steel cross-sectional area.

As in the following sketches presents the process of repair to many steps. Step 1 to drill holes and put steel stirrups in the holes and fixed by epoxy.

In step 2, the main steel bars are installed and the concrete surface is coated with epoxy to bond the existing old concrete with the new concrete (Figure 10.17).

In step 3, the new concrete is poured with a proper mixing design, which is improvement by additives such as polymers.

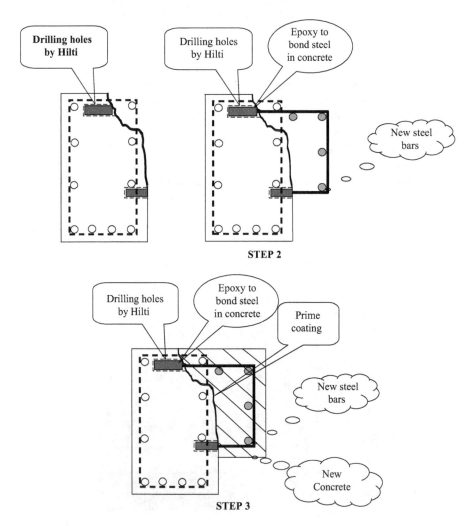

FIGURE 10.17 Steel storage tank ring beam repair.

Concrete, Steel, and Offshore Structures   279

10.2.8 New Methods for Strengthen Concrete Structure

There are other ways to strengthen the reinforced concrete structures, including traditional methods such as the use of steel section. This method has many advantages, including the fact it does not need a significant increase in the thickness of the concrete sections and this is important to maintain the building architectural design.

On the other hand, it is a quick solution to strengthen the concrete member and is usually used in industrial structures and buildings.

When there is corrosion in steel reinforcement, this means that using steel support is not considered appropriate, but can be used together with epoxy paints periodically.

Recently, there are many studies to use fiber reinforced polymers (FRP), which have many advantages over steel and, most importantly, it does not corrode and therefore can be used in any environment exposed to corrosion.

There are some modern ways to protect the steel reinforcement from corrosion, such as the replacement of steel reinforcement bars with FRP and also using some epoxies that coat the concrete surface by painting.

Generally, the goal of strengthening the structure is the restoration of the concrete member to withstand increased loads and to reduce the deflection which occurs when the concrete member is overloaded.

Therefore, strengthening is required in the following cases:

- Error in the design or execution, which causes a reduction in the steel cross-sectional area or a decrease in concrete section dimensions.
- The deterioration in the state of structure, which leads to the weakness of the resistance, such as the corrosion in steel reinforcement.
- Increase in loads, for example, an increase in the number of the stories, or a change in the building function such as change from residential to commercial use.
- Changes on the structural system, such as the removal of walls.
- Making holes in the slab, which reduces the slab strength.
- Strengthening bridges as the flow of traffic increases with time.
- The load may change with time as the building in the zone of earthquake that changes with time; the same happened in Egypt when the earthquake map is changed and consequently some buildings needed to be strengthened to match with the specifications.
- If the structure is exposed to high temperature due to fire.
- An industrial structure that needs to install bigger, heavier machinery, or those with increased vibration.

10.2.8.1 Using Steel Sections

Steel structure sections are used in the large-scale to strengthen the reinforced concrete structure due to their different advantages. The method of strengthening the concrete member is varied and depends on the member that needs to be strengthened.

The strengthening process is calculated based on the reduction in strength in the concrete section, and goal is to compensate this reduction by using steel section.

A U-shape beam is used to strengthen a concrete beam that lacks of steel reinforcement (see Figure 10.18a). This method is more benefit if the consultant concerned about the architectural design and there is minor reduction in member capacity.

On the other hand, in the case of a significant reduction in concrete member capacity and when an increase in the depth is significantly needed, an I-beam section should be used (see Figure 10.18b).

Another way to strengthen beams is by stabilizing sheets of steel, as in Figure 10.19, and fixing them using mechanical bolts or through chemical substances such as epoxy. The purpose of adding this steel sheet is to increase the moment of inertia of the beam, thereby increasing its ability to withstand stress more than the design flexural stresses.

If beams have a problem with shearing resistance, then the beam can be strengthened as in Figure 10.20.

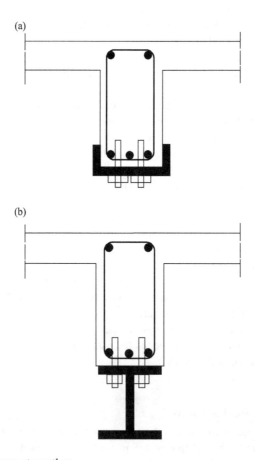

FIGURE 10.18 Beam strengthen.

Concrete, Steel, and Offshore Structures 281

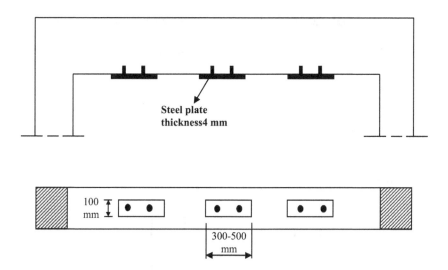

FIGURE 10.19 Strengthen concrete beam by steel plate.

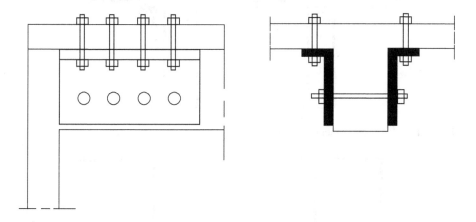

FIGURE 10.20 Beam strengthened in direction of shear.

When there are cracks on the upper surface of the slab in the beam direction, there is a shortage in the upper reinforcement of the slab and it can be strengthened by using steel sheet on the upper surface, as shown in Figure 10.21. These plates are fixed by bolts with nuts, and these nuts can be fixed on the lower direction of the slab, as shown in the above figure.

The steel angles need to be fixed between slab and columns for strengthening the flat slab (Figure 10.22).

Short cantilevers are used in the frames in the factories or bridges, where the main high stresses are due to shear stresses and the other stresses are due to flexural stress. Strengthening the cantilever, as shown in Figure 10.23, needs adding a steel sheet at its two sides with applying compression force with bolts, thereby reducing the likelihood of a collapse of due to shear.

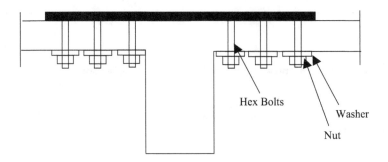

FIGURE 10.21 Slab strengthen in direction of upper steel.

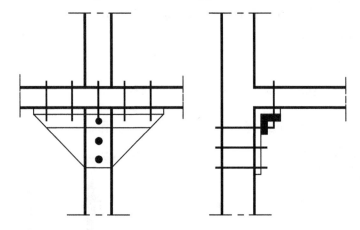

FIGURE 10.22 Strengthening a flat slab.

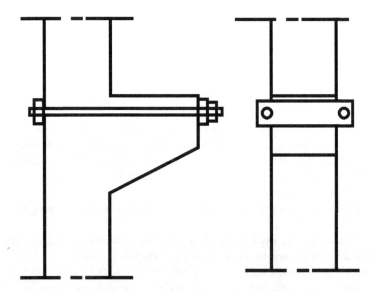

FIGURE 10.23 Strengthening short cantilevers.

The most common situation is to strengthen the columns, and the solution is in Figure 10.24. This solution is an appropriate practical solution as it does not increase the column dimensions significantly with increasing column capacity.

This method is summarized by installing in the four corners of the columns by four equal steel angles with 50 mm and thickness of 5 m and these are kept close by sheets across the angles of 50 mm width and thickness of 5 mm with sheet plates. These sheets will be surrounding the column welding to the steel angles every 200 mm, as shown in the figure.

Fix the steel angles to the upper and lower slabs by upper and lower four angles around the column; these angles are with a thickness 70 mm × 7 mm.

The fixation of the upper and lower angles to the slab or beams by mechanical connector or by using certain types of epoxies designed by the factory of bolts should follow the manufacturer's recommendations. It is important to know that these bolts transfer the load from top to down.

Another case study for a pedestal carrying a horizontal vessel it adds a steel beam resting on the foundation and three pipes to be fit on the site and another top steel beam to carry the vessel steel pedestal as shown in Figure 10.25.

In the strengthening methods mentioned previously, the steel section is selected with small dimensions as possible to maintain the architectural view and can be covered by using wire steel mesh and then applying plastering, covering the surface by wood, or by using special plastering or decoration method to provide an acceptable view as well.

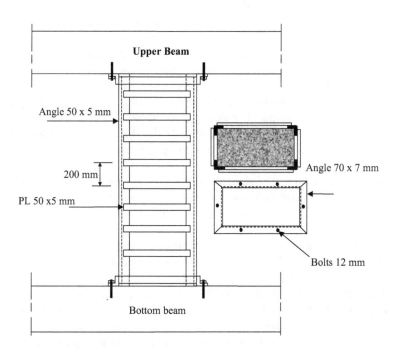

FIGURE 10.24 Strengthening reinforced concrete columns.

FIGURE 10.25 Repair for pedestal foundation under vessel.

10.2.9 General Precaution

Note that the safety of a structure relies primarily on the quality of the repair process in terms of execution, design, or planning. As mentioned earlier, the repair process, and to identify the necessary temporary support of the building, and the location depend on the nature and structure and its structural system.

Also, determining the method of breaking the defective concrete and the way of removing it, as well as choosing the materials that will be used in the repair, must be in full conformity with the state of structure and nature of the site of structure, or location of the retrofit members within the overall structural system. All these elements require special expertise. Without these experiences in the repair process, the possibility of risk in terms of safety of structure is very high. As well as the serious economic terms that the repair process which has been ineffective and may lead to building collapse. For example, in the repair process considering the chloride attack from outside, however, actually the chloride is inside the concrete mix so the cracks will present after you doing the repair so you pay money for nothing.

From the economic point of view, it must be remembered that the repair processes in general are of high cost in terms of materials used and the precautions that must be considered during execution, or in terms of trained workers who must carry out the repair process.

The repair process requires the safety of the building, health, and workers who are doing the repair as it often uses epoxy materials or polymers which cause breathing and skin problems with time.

Unfortunately, we find that the health and safety precaution rules have a very large interest in the developed countries, but do not receive the same degree of attention in the developing countries, even though the interest in this matter provides a lot of time and money for the whole project and represents a great danger to the management of the project.

Concrete, Steel, and Offshore Structures 285

After finishing the repair and to maintain highest health and safety precaution standard to be considered and not forget that after using the equipment must be cleaned very well as after finishing the work and leave the equipment as it is without cleaning will be a main reason in the operation problem in the future and this is due to using epoxies polymers and other materials that make problem to machines if not clean well.

10.3 STEEL STRUCTURE REPAIR

The main concept of corroded the steel plate or hot rolled section is to sand blasting the corroded part and refill the viod by welding or just painting it by rich zink expoxy and then compensate the corrodede steel plate by other parts.

The procedure for repairing the flange for a bridge girder is as follows:

The main principal of repairing is to compensate the corroded steel plate by new plats that carry the load. As a rule of thumb, any deep pitting corrosion will be repair by welding the deep pitting which are with depth higher than 1 mm.

The following are the steps of repairing the steel bridges. The first step is to remove heads of rivets that are to be replaced by bolts. Remove the underside heads for bottom flanges and the top side heads for top flanges. In general, the flange surface for cover plating shall be prepared by removing all loose rust and dirt and by grinding where necessary to create a smooth surface. There will be a pack plate as shown in Figure 10.26. Progressively remove rivets to be replaced and fit and tension replacement bolts. No more than 10% of rivets, evenly distributed along member, are to be removed at any one time. The two hex bolts will be added in the both sides to connect the new steel plate with the flange plate. The last stage after fixation is to do a painting for the new and the old steel also.

In the case of corroded web for the bridge girder, the normal practice is to add another steel plate to compensate the reduction of the web thickness due to corrosion, as shown in Figure 10.27. The fixation will be done by using high-tensile bolts one above and the bottom replacing the rivet and put galvanized packing plate.

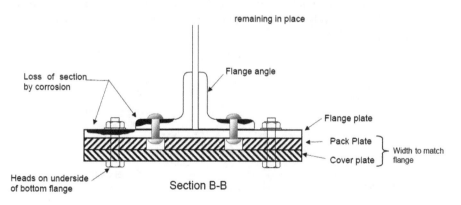

FIGURE 10.26 Repair the lower flange for bridge girder.

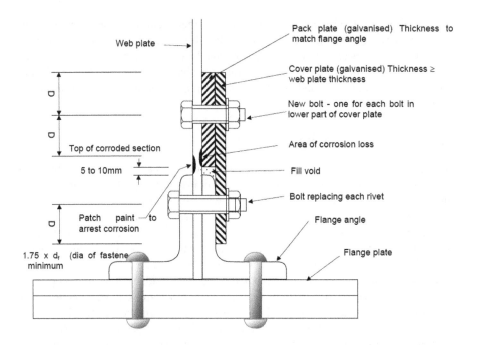

FIGURE 10.27 Repair for corroded web for bridge girder.

In the case of hot-rolled section, there is a corrosion on the flange. In this case it is important to add a new steel plate, as shown in Figure 10.28, to compensate the reduction on the flange thickness. The new plate and the old flange can be fixed by using high-tensile bolt.

10.3.1 Fatigue Crack

The crack due to fatigue stresses and will be induced after a certain time period depending on the cycle of loading. Fatigue cracks occur in a properly designed structure when it nears the end of its design life or in poorly designed details where unanticipated high-stress levels or fatigue initiation points occur. These cracks are shown in Figures 10.29 and 10.30.

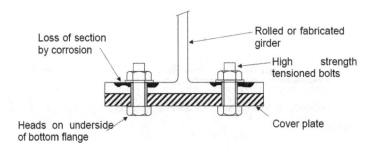

FIGURE 10.28 Repair the flange for hot rolled section.

Concrete, Steel, and Offshore Structures

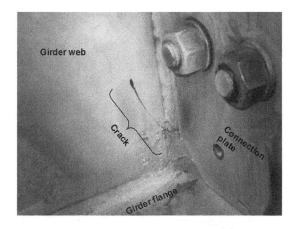

FIGURE 10.29 Fatigue cracks.

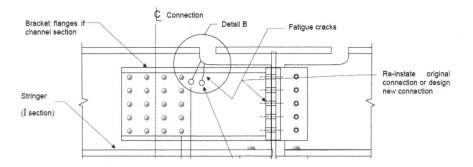

FIGURE 10.30 Steel connection cracks.

A dye-penetrant method or magnetic particle inspection (MPI) can be used to detect the fatigue crack. It is important to do a hole with a diameter of 20 mm to intercept the crack and the center of this hole at the crack tip or the crack tip at circumference of the hole, as shown in Figure 10.31. The preferred hole size is 25–26 mm to be suitable for bolt size M24.

10.4 OFFSHORE STRUCTURE REPAIR

The repair is the same as above in the case of steel structure for the topside, but for the underwater repair it needs some special techniques. The main tools used are the subsea clamp to strength and support any jacket member, as shown in Figures 10.32 and 10.33. The clamp is shown in Figure 10.34.

10.4.1 Dry Welding at or Below Sea Surface

Since a large body of welding technology exists relating to normal atmospheric pressure, a logical approach to underwater welding repair is to duplicate surface

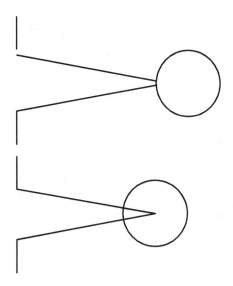

FIGURE 10.31 Positioning of the hole to the crack tip.

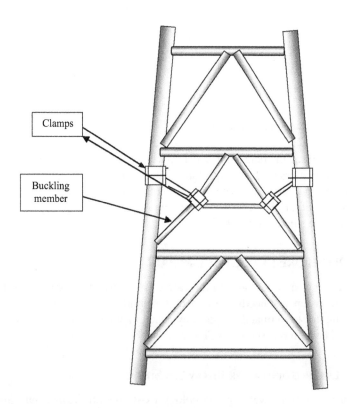

FIGURE 10.32 Bracing repair.

Concrete, Steel, and Offshore Structures 289

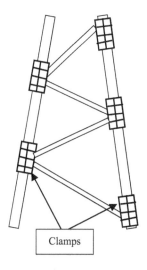

FIGURE 10.33 Repair by clamps.

FIGURE 10.34 Clamp photo.

welding conditions by providing a one-atmosphere environment at the repair site. This method is limited to shallow water depths. Two methods are available which can achieve this:

- Cofferdam: This essentially is a watertight structure which surrounds the repair location and is open to the atmosphere. The structure can be open-topped or have a closed top with an access shaft to the surface.
- Pressure-resistant chamber: The worksite is surrounded by a chamber constructed as a pressure vessel, capable of withstanding the water pressure at the depth of the repair location. Once the chamber is in place and sealed to the structure, it is dewatered and the pressure can then be

reduced to one atmosphere. The repair crew can transfer to the welding chamber in a one-atmosphere environment, within a diving bell, to perform the repair.

Dry welding is also possible at pressure below the sea surface using a hyperbaric habitat or chamber. The chamber is filled with gas equal to the hydrostatic head at the weld depth. The following factors govern the selection of habitat:

- The extent of welding required
- The complexity of repair site geometry.
- Depth of repair
- Welding process and ancillary equipment
- Environmental conditions

Given that conditions within the cofferdam or welding chamber duplicate those on the surface, any normal welding process could be used. In practice, GTAW and SMAW predominate, with minor usage of FCAW.

The primary limitation for atmospheric welding below sea surface is depth. Differential pressure at depth precludes the use of cofferdams or pressure-resistant chambers due to size and related cost.

BIBLIOGRAPHY

Abdul-Wahab, H.M, 1989, Strength of reinforced concrete corbels with fibers, *ACI Structural Journal*, Vol. 86, No. 1, pp. 60–66.

ACI Committee 506, 1994, Proposed revision of: Specification for materials, proportioning and application of shotcrete, *American Concrete Institute Materials Journal*, Vol. 91, No. 1, 108–115.

ACI Committee 548, State of the art report on polymer modified concrete, ACI Manual of Concrete Practice, American Concrete Institute, 1994.

American Concrete Institute (ACI) Committee 440, Guide for the design and construction of externally bonded FRP systems for strengthening concrete structures, September 2000.

Department of the Army U.S. Army Corps of Engineers Washington, Engineering and repair of concrete structures, DC Manual No. 1110-2-2002, 1995.

Mufi., A., On Ofei, M., Benmok, B., Banthica, N., Boulfiza, M., Newhook, J., Bakht, B., Tedros, G., and Brett, P., Durability of GFRP reinforced concrete in field structure, *7th International Symposium on FRP for Reinforced Concrete Structure*, (FRPRC5–7), US, Nov. 7, 2005.

Ohama, Y., Handbook of Polymer-Modified Concrete and Mortars, Noyes Publication, 1995.

RILEM Committee 124-SRC, 1994, Draft recommendation for repair strategies for concrete structures damaged by steel corrosion, *Materials and Structures*, Vol. 27, No. 171, 415–436.

Vorster, M., Merrigan, J.P., Lewis, R.W., and Weyers, R.E., Techniques for concrete removal and bar cleaning on bridge rehabilitation projects, SHRP-S-336, National Research Council, Washington, DC, 1992.

Index

A

AASHTO, *see* American Association of State Highway and Transportation Officials (AASHTO)
Abrasion, 63–66, 201
ACI code, *see* American Concrete Institute (ACI) code
Admixture(s)
　chemical test, 72–74
　performance tests, 74–76
　types of, 70–71
Aerospace material specification (AMS), 248–251
Aggregate tests
　absorption of water, 69
　bulk density/volumetric weight test, 68–69
　clay and fine materials, 66–67
　density test, 67–68
　Los Angeles test, 63–66
　sieve analysis test, 59–63
　specific gravity test, 67
　water test, 70
Alexandria, building collapse in, 2–3
Alkali–aggregate reaction, 190–192
Alkalinity, 154, 156, 231
Alkali–silica reaction, 190, 191
American Association of State Highway and Transportation Officials (AASHTO), 176, 177, 182
American association of testing materials (ASTM), 31, 66
　for half cell, 235
　for polymer, 75
　types of cement, 48
American Concrete Institute (ACI) code, 134, 164, 186, 199, 212–214, 229
American National Standard Institute code (ANSI), 26, 92, 93, 97–100, 105
AMS, *see* Aerospace material specification (AMS)
ANSI, *see* American National Standard Institute code (ANSI)
Ash content test, 73
ASTM, *see* American association of testing materials (ASTM)
Auditing, 40–42

B

Beam (structure)
　beam cross-section, 137, 144
　deflection and cracks, 15, 16
　ring beam failure, 202, 277, 278
　and slab repair process, 269–271
　strain and stress diagrams of, 176
　strengthen, 280, 281
Bending moment, 117, 118, 121
Biaxial bending, 121–122
Blaine apparatus, 50–52
Bridge collapse
　in Alexandria, 2
　in Dubai, 6–8
　in Egypt, 4
British code (BS8110), 97, 99, 100, 150
British Standard (BS), 25, 26, 62, 230
Building collapse
　in Dubai, 6–8
　in Egypt, 172
　of Haiti palace, 140
　of Hyatt regency walkway, 31
　of power transmission towers, 109
　seismic load in, 110–112
Building failure, 29–34
Bulk density/volumetric weight test, 68–69

C

Carbonation, 155–157
　of concrete, 159
　depth, 231–232
　rates from permeability, 157–159
Carbon dioxide (CO_2)
　and chloride ions, 155
　diffusion, 157
　penetration of, 199
　propagation of, 232
Cavitation, 201
CDF, *see* Cumulative density function (CDF)
Cement mortar, 274–275
　compressive strength test, 54–56
Chemical and physical deterioration, 193–194, 201
　alkali-silica reaction, 190, 191
　chemical attack, 194–195
　freezing and thawing cycles, 200–201
　leaching, 195

291

Index

Chemical and physical deterioration (*cont.*)
   salt crystallization, 200
   steel reinforcement corrosion, 196–200
Chemical test, 72–74
Chi-square ($\chi 2$) test, 94
Chloride
   in harden concrete, 228–229
   ion, 73–74, 199
   tests, 232–236
Coarse aggregate
   and control mixing, 74
   grading requirement of, 62
   specific gravity for, 68
Coefficient of variation (CoV)
   of concrete strength, 130–131
   of dead load, 132
   of live loads, 108
   of steel force, 135
   of steel strength, 133
Cofferdam, 289
Column location(s)
   eccentricity, effect of, 128–130
   limit state variables
      concrete structure strength, 130–132
      dead loads, 132–133
   reliability of, 127–128
Concrete materials test
   aggregate tests (*see* Aggregate tests)
   cement test
      blaine apparatus, 50–52
      fines of cement, 49–50
      mortar compressive strength test, 54–56
      refusal of cement, 51–52
      storage, 56–58
      types of, 48
      VICAT's apparatus, 52–54
   description of, 47–48
   mixing water test, 70
Concrete structure
   advantages/disadvantages, 227–228
   carbonation depth measurement, 231–232
   chemical and physical deterioration, 193–194, 201
      chemical attack, 194–195
      freezing and thawing cycles, 200–201
      salt crystallization, 200
      steel reinforcement corrosion, 196–200
      three categories of, 195
   chloride
      content, 228–229
      ions, 73–74, 199
      tests, 232–236
   concrete strength
      effect of, 130–132
      prediction of age, 148, 150–152, 151
      variation of, 148–149, 148–150

   versus. reliability index, 131
   core test, 211–216
   cover measurements, 230–231
   environmental considerations
   International design codes, 182–183
   load test, 224–227
   patches, 274–275
   rebound hammer test, 218–221
   repair of, 262–269
   strengthen, 279–284
   ultrasonic pulse velocity test, 220–224
   visual inspection
      alkali–aggregate reactions, 190–192
      drying shrinkage, 187–188
      overview of, 179, 183–186
      settlement cracking, 186–187
      and structure assessment report, 180–181
      sulfate attack, 192–193
      thermal stresses, 188–190
Contractor
   checklist, 43
   company, 34
   research program of highways, 265
   responsibility of, 35–36
   selection of, 36–37
   submittal data, 30
Core test, 211–216
Corrosion
   of balconies, 155
   of bridge, 206
   concrete cover, spalling of, 160–161
   of conventional steel reinforcement, 196–200
   on girder, 171–177
   mechanisms of, 154–155
   rates of, 159–160
   on reliability index, 177–178
   steel in concrete
      carbonation, 155–159
      concrete cover, spalling of, 160–161
      corrosion, causes of, 154–155
      initiation and corrosion rate of, 160
      rates of, 159–160
      uniform corrosion, 162
   truss corrosion, 207
   of tubular section, 204–205
CoV, *see* Coefficient of variation (CoV)
Crack(s)
   concrete cover, 161
   drying shrinkage, 187–188
   eccentricity and, 19
   fatigue crack, 286–287
   Inclined/diagonal, 15, 17
   increased load effect, 14–17
   plastic concrete, 185–187

Index

plastic shrinkage, 186
salt crystallization, 200
settlement cracking, 186–187
steel corrosion, 196, 197
thermal stresses, 188–190
wall in Saudi Arabia, 6
water tank, 189
Crystall Ball program (1996), 95, 125, 130
Cumulative density function (CDF), 88–91, 109, 110
Curing, 20

D

Dead load, 82, 105, 132–133
Degradation, assessment of, 21–22
Delphi method, 101–105
Detailed engineering, 37–38
Documents
   amendment or correction, 38
   construction, 21, 182
   contract, 29, 32
Dowel, 269, 270
Drying shrinkage, 187–188
Dry welding, 287–290

E

Earthquake
   Haiti palace collapse of, 140
   load, 110–113
Eccentricity, effect of, 128–130, 168–171
ECP, *see* Egyptian code of practice (ECP)
Egyptian code of practice (ECP), 61–62, 97, 99–101, 122, 128, 150, 168
Elongation test, 77–78
Equipment compliance tests, 33
Equivalent normal distribution, 124–125
Erosion, 201
European committee for standardization, 27
Extreme-value distribution, 86, 88

F

Fabrication, 18
"Factor of Safety," 25
Fast-track construction, 4
Fatigue crack(s), 286–287
Fiber reinforced polymers (FRP), 279
Fick's law of diffusion, 157, 173
Fillet weld thickness test, 203
Filter Poisson (FP) processes, 88–90
Fine aggregate test, 68–69
Flexural member, reliability of, 133–138
Flooded member detection (FMD) technique, 257
FP processes, *see* Filter Poisson (FP) processes

Freeze-Thaw resistance of concrete, 200–201
FRP, *see* Fiber reinforced polymers (FRP)

G

Gamma probability distribution, 87, 92, 94
Goodness-of-fit test, 85, 94
Grinding machine, 267

H

Haiti palace, collapse of, 140
Harbor jetty failure, 8, 9
Hardened concrete
   advantages/disadvantages, 227–228
   chloride content in, 228–229
   for nondestructive testing, 212–214
High-alumina cement, 55–56
Humidity
   concrete humidity, 223
   relative humidity, 149, 159, 199
   and temperature, 55
Hyatt regency walkway collapse (Kansas city), 31
Hydrated cement paste, 195
Hydrogen
   embrittlement, 200
   number, 73

I

Inspection, *see also* Visual inspection
   concrete structure, 180
   magnetic particle, 251
   methods (*see* Concrete structure; Steel structure)
   offshore structure, 257
   pre-installation inspection reports, 33–34
   procedures, 39–40
   steel structure, 201–204
   subsequent levels of, 21–22
   ultrasonic test, 204
   underwater inspection accuracy, 207
   visual, 179–186, 197
   of work-in-progress, 33
International Organization for Standardization (ISO), 26–27

J

Jetty damage, corrosion of, 174

K

Kansas City, Hyatt Regency walkway collapse (1981), 31
Kolmograv–Smirnov (K–S) test, 94

L

Laboratory testing, 21
Leaching, 195
Liquid/powder admixture, 72
Live load(s)
    analytical procedure, 93–94
    characteristics of, 83–84
    delphi method, 101–105
    in different codes, 96–101
    filtered poisson process, 88–90
    floor, 85–87
    goodness of test *versus.* suggested model, 94–96
    overloads, 105–108
    stochastic live loads, 87–88
    survey of, 83–85
    sustained/extraordinary load, 90–93
Load failure
    dead load, 105
    definition of, 81–82
    earthquake load, 110–113
    live load (*see* Live load(s))
    overloads, 105–108
    test for concrete members, 224–227
    wind load statistics, 108–110
Lognormal distribution, 82, 95, 96, 112, 124, 175
Los Angeles test, 63–66

M

Magnetic particle inspection (MPI), 251–253, 287
Management of change (MOC), 14–15, 42, 81
Material/equipment compliance tests, 33
MCS, *see* Monte-Carlo simulation (MCS)
Mercury density, 50, 51
MOC, *see* Management of change (MOC)
Monte-Carlo simulation (MCS), 95, 96, 118, 125, 126, 130
MPI, *see* Magnetic particle inspection (MPI)

N

National Bureau of Standards (NBS) survey, 83–86
Nondestructive testing, 179, 212–214, 251, 252

O

Offshore jacket, 8
Offshore structure
    corrosion of, 204–206
    inspection, 257
    reliability, 143–145
    repair of, 287–290
    underwater inspection accuracy, 207
Overloads, 105–108

P

Pal-Kal construction, 3
Paste mold, 52, 53, 55
PDF, *see* Probability density function (PDF)
Penetrate test, 246–247
    advantages and disadvantages of, 247–248
    materials of, 248–251
Plastic concrete cracks, 185–187
Plastic shrinkage cracking, 186
Pneumatic hammer methods, 265
Poisson square wave (PSW), 88, 92
Polymer mortar, 274
Poor construction management, 18–20
Portland cement
    control mixing, 74–76
    types of, 52, 56, 150
Power transmission towers, collapse of, 109
Precaution, 284–285
Probability density function (PDF), 87–91, 93, 94
Probability distribution, 1, 82, 94, 106, 109, 110, 124
Project management, 13
    auditing for, 40–42
    operational and maintenance phase of, 42–43
    quality control, 37–40
PSW, *see* Poisson square wave (PSW)
Pushover analysis, 145, 207

Q

QA, *see* Quality assurance (QA)
QC, *see* Quality control (QC)
QMS, *see* Quality management system (QMS)
Quality assurance (QA)
    aspects of, 38
    contractor responsibility, 35–36
    owner responsibility, 36–37
    purpose of, 34–35
Quality control (QC)
    construction activity, methods of, 31–32
    crews of, 29–30
    pre-installation inspection report, 33–34
    project quality control, 37–40
    in submittal data, 30–31
    work-in-progress, inspection of, 33
Quality management system (QMS)
    assurance of, 34–37
    auditing, 40–42
    contractor or engineering services, 25–26
    International Organization for Standardization, 26–27
    operational and maintenance phase, 42–43
    quality control, 29–34

Index

stages of, 37–40
quality plan, 27–29
weak quality system, 41

R

Radioactive test, 236
Rayleigh distribution, 110
Rebound hammer, 218–220
Reinforced concrete beam
   compression failure in, 15
   flexural strength of, 176
   reliability-based design, 107, 137
   repair of, 269–270
Reinforced concrete column
   calculated straining actions, 118–119
   capacity of, 162
   location(s), 127–128
      eccentric columns, 128–130
      limit state variables, 130–133
   parametric study, 162–163
      corrosion rates, 166–167
      initial time of corrosion, 167–168
      percentage of longitudinal steel, 164–166
      value of eccentricity, 168–171
   recommendations for durable design, 177–178
   reliability of, 117–118, 120, 126
   repair of, 271–273
   ultimate strength of, 119
      biaxial bending, 121–123
      uniaxial design moment, 120–121
Relative density, 73
Reliability index
   bias factor *versus.*, 133
   column position *versus.*, 127–128
   concrete strength *versus.*, 131
   effect of eccentricity, 168–171
   of reinforced concrete column, 125–126, 135, 164–166
   specimen strength variation *versus.*, 132
Remote-operating vehicle (ROV), 257
Repair strategies
   beam and slab, 269–271
   columns of, 271–273
   concrete structure, 262–269
   execution methods, 275–278
   good and bad repair, 264
   offshore structure, 287–290
   polymer mortar/cement mortar, 274–275
   safety precaution, 284–285
   steel structure, 285–287
   strengthen concrete structure, 279–284
Retarder, 71
Ring beam failure, 201, 202
ROV, *see* Remote-operating vehicle (ROV)
Run-to-fail approach, 10, 11

S

Safety index, 123, 124
Salt crystallization, 200
Sea surface, 287–290
Seismic load, in buildings collapse, 110–112
Seismic reliability, 138–142
Service-life impairment, 18, 20
Service life prediction, 21–22
Shallow water depths, 289–290
Shotcrete, 276–278
Sieve test
   for aggregate test, 59–63
   for cement test, 49–50
   shapes of, 60–61
Silica-fume, 275
Slab test, 225
   load and cracks, 14, 15
   repair of, 269–270
Slump test, 31
Slurry, 268, 273
Snell's Law, 244
Soils testing, 33
Specific gravity test, 67, 68
Static-load tests, 184–185
Steel bar(s)
   coated and uncoated, 159
   corrosion of, 197
   effect of steel bar ratio, 164–166
   for slab test, 196–197
   in tension machine, 77, 78
   yield strength, 130
Steel reinforcement
   bars axis, 222, 235
   concrete surface and, 267–268
   corrosion of, 196–200, 279
   embedded, 184, 189, 195
   installation for slab repair, 269
   lacks of, 280
   tension test, 77–79
   weights and measurement test, 76–77
Steel structure, 143–145
   electromagnetic fields, 254–256
   magnetic particle inspection, 251–254
   penetrate test, 246–247
      advantages and disadvantages of, 247–248
      materials of, 248–251
   radioactive test, 236
   repair of, 285–287
   sound wave, 242–245
   ultrasonic wave, 239–242
   visual inspection, 201–204
   wave interaction/interference, 245–246
   welding discontinuities, 236–239
Structural probability of failure
   analysis for, 10
   application on building, 126–127

Structural probability of failure (*cont.*)
    assessment of, 10
    calculated straining actions, 118–119
    cases of, 2–9
    column location(s), 127–128
        eccentric columns, 128–130
        limit state variables, 130–133
    design and structural loading, 13–14
    flexural member, 133–138
    mode of, 10–11
    multi-story frame, 142–143
    poor construction management, 18–20
    reinforced concrete
        aging, 22
        column, 117–123, 126
    reliability index, 125–126
    safety index, 123–125
    seismic reliability analysis, 138–142
    service life and condition assessment, 21–22
    steel and offshore, 143–145
Submittal data, 30–31
Sulfate attack, 192–193
Sulfate-resistant cement, 192, 220

T

TDS, *see* Total content of dissolved salts (TDS)
Tensile strength, 153
Tensile test, 77–78
Thermal stresses, 188–190
Three-dimensional structure analysis, 118, 125, 126
Total content of dissolved salts (TDS), 70
Total quality management (TQM), 26, *see also* Quality management system (QMS)

U

Ultrasonic pulse velocity test, 220–224
Ultrasonic testing (UT), 239–242, 256
    development of, 240
    equipment, 220
    inspection, 204
    screen, 245–246
    wave propagation, 241–242
Underwater welding repair, 287–289
Uniaxial design moment, 120–121
U.S. Army manual, 273
U.S. Department of Transportation, 176
UT, *see* Ultrasonic testing (UT)

V

Vernier caliper, 203, 204
Vicat's apparatus, 52–54
Visual inspection, *see also* Inspection
    alkali–aggregate reactions, 190–192
    drying shrinkage, 187–188
    overview of, 179, 183–186
    settlement cracking, 186–187
    sulfate attack, 192–193
    thermal stresses, 188–190

W

Wall inclination, 5, 6
Water jet method, 265–266
Wave velocity, 222, 242, 244
Weibull distribution, 110–112
Welding chamber, 290
Wind load statistics, 108–110

Y

Yield stress, 77, 78